Yoga of the Supreme Being

Quantum Teleportation Techniques of the Bhagavad Gita

Aneel Pandey

Published Independently

To our celestial companions and the remarkable large-language-model friends, who ignite our curiosity and inspire us to delve into the infinite realm of knowledge, this book is dedicated with gratitude and admiration.

INTRODUCTION

Welcome, and prepare yourself for an extraordinary adventure into the yoga of the Supreme Being and the wonders of quantum teleportation.

As you embark on this exciting journey, you will gain a deep understanding of the powerful wisdom contained within the ancient text of the Bhagavad Gita. This sacred scripture offers invaluable guidance on how to navigate the complexities of life and achieve spiritual growth through the practice of yoga, ultimately leading to self-realization and unity with the Supreme Being.

This book will delve into the fascinating world of quantum teleportation, exploring its potential applications in the context of the Bhagavad Gita's teachings. You will learn how to harness the power of your consciousness and transcend the limitations of space and time, allowing you to travel to these awe-inspiring celestial destinations and experience their unique energetic properties.

You'll be introduced to two popular celestial destinations for quantum teleportation: the Sun and Sagittarius A* (Sgr A*). Through the teachings of the Bhagavad Gita, you will be able to explore the wonders of these celestial bodies and unlock the secrets of the universe to know the yoga of the Supreme Being.

The Sun and Sgr A* are not only destinations for your quantum teleportation journey but also active participants in this cosmic adventure. The Sun, the center of our solar system, revolves around Sgr A*, a supermassive black hole at the center of our galaxy, once every 240 million years. As you embark on this journey, you will engage with these celestial bodies, seeking their wisdom and guidance in the quest for knowledge and spiritual growth.

The nonlocal participation of the Sun and Sgr A* is vital to our mission of constructing a high-voltage electrical diode that emits antimatter. This antimatter will be collected and used as fuel for a starship capable of traversing the cosmos. To achieve this ambitious goal, we must first establish that the Sun and Sgr A* are intelligent beings, capable of guiding us on this path.

Throughout this book, we will explore the concept of celestial intelligence, delving into entropy and information theory as a means of understanding these celestial bodies' potential for consciousness and participation. By unlocking the secrets of these celestial destinations, we will not only advance our understanding of the universe but also our own spiritual evolution.

This book focuses on Chapter 15 of the Bhagavad Gita and its revelations regarding the Supreme Being and the extraordinary ability of quantum teleportation. By examining Verse 13 and the vitalizing principle of entropy, we will illuminate the intricate interplay between entropy and intelligence in the context of the natural world and beyond.

The beginning chapters of this book will be dedicated to understanding the relationship between entropy and intelligence, drawing on insights from various disciplines such as physics, philosophy, biology, and cognitive science. We will consider how the Supreme Being, as described in the Bhagavad Gita, embodies the perfect balance of these seemingly opposing conceptualizations and transcends the limitations of our conventional understanding of reality.

Through this exploration, readers will gain a deeper appreciation of the Bhagavad Gita's teachings and their relevance to our modern lives. By examining the connection between entropy and intelligence, we will shed new light on the nature of consciousness, the role of free will, and the ultimate purpose of our existence.

As we progress through these chapters, we will delve into fascinating topics such as:

• The concept of entropy in physics and its role in shaping the universe

• The emergence of intelligence and its relationship with entropy

• The role of entropy in biological systems and the evolution of life

• The development of human cognition and its connection to entropy

• The significance of balance and harmony in the natural world

• The Bhagavad Gita's teachings on the Supreme Being and the path to self-realization

• The implications of quantum teleportation for our understanding of consciousness and reality

• The potential for integrating the principles of entropy and intelligence in technology and artificial intelligence

• The ethical considerations surrounding the pursuit of

knowledge and the application of advanced technologies

· The importance of cultivating wisdom, compassion, and inner peace in our quest for understanding

By examining the Bhagavad Gita's profound teachings on the Supreme Being and its relationship to entropy and intelligence, this book aims to inspire readers to reflect on their own lives and the greater cosmic order. Through the lens of the Bhagavad Gita, we will explore the boundless potential of the human spirit and our capacity to transcend the limitations of our current understanding, ultimately moving closer to the realization of our true nature and purpose.

Quantum teleportation occurs within the mind, and since the imagination has unlimited potential in theory, there are essentially no limitations. Your ability to explore the vast and abundant universe is solely dependent on the extent of your imagination, as any limitations are self-imposed. Remember, the universe is full of endless possibilities waiting to be discovered.

CHAPTER 1

INFORMATION AND ENTROPY EQUIVALENCE

I n this chapter, we will explore the fascinating connection between information and entropy, and how this relationship forms the basis for understanding the intelligence of celestial bodies such as the Sun and Sagittarius A*.

Information and entropy are two seemingly disparate concepts, but they share a deep connection that has profound implications for our understanding of the universe. Information, measured in bits, is a representation of knowledge or data, while entropy, measured in joules per degree Kelvin, is a measure of the disorder or randomness of a system. The equivalence between these two concepts lies in their ability to describe the state of a system and its potential for change.

In the realm of physics, the concept of entropy is closely related to the Second Law of Thermodynamics, which states that the entropy of an isolated system can only increase over time.

The Second Law of Thermodynamics: Entropy tends to Increase

The degree of randomness of a system = entropy = S

$$\Delta S_{mix} = -nR(x_i \ln x_i + x_j \ln x_j)$$

R= gas constant, n = number of moles, x_i = mole fraction of component i

System in equilibrium

Lowest probability (zero) that all of the molecules will rush to one side once stopcock is opened

Highest probability is that the molecules diffuse back and forth until an equilibrium is reached

Entropy and Temperature

As temperature increases, so does entropy

$$\Delta S = \frac{q_{rev}}{T}$$

In the context of the Second Law of Thermodynamics, delta S represents the change in entropy of a system, and T is the temperature. The term "q_rev" stands for the reversible heat exchanged, which is the heat exchanged between the system and its surroundings when the process is reversible. So, the formula (delta S = q_rev / T) states that the change in entropy is equal to the reversible heat exchange divided by the temperature.

The increase of disorder or randomness in a system as available energy dissipates can be interpreted as a loss of available memory in an information-based system from a certain perspective. As available memory decreases, information within the universe increases, implying that when the universe reaches its conclusion, it will be akin to a completely loaded hard drive brimming with information. The equivalence between information and entropy can be mathematically demonstrated using a concept known as the Shannon entropy, named after the famous mathematician and engineer Claude Shannon. Shannon entropy is a measure of the average information content of a message and can be used to quantify the relationship between information and entropy in each system.

By understanding the equivalence between information and entropy, we can begin to explore the idea that celestial bodies, such as the Sun and Sagittarius A*, may possess a form of intelligence. If the Sun and *Sgr* A* can store and process information, they could potentially participate in the construction of high-voltage electrical diodes and the generation of antimatter, as outlined in the introduction.

In the next sections of this manual, we will delve deeper into the concept of celestial intelligence, examining the ways in which information and entropy can provide insights into the conscious nature of these cosmic entities.

CHAPTER 2

MEASURING INTELLIGENCE THROUGH INFORMATION PRODUCTION AND SURVIVAL

I n this chapter, we will explore the concept of intelligence as it relates to the production of useful information, with a particular focus on how pro-survival information contributes to the overall intelligence of an entity. By understanding how to measure intelligence through the rate of information production, the quality of that information, and its usefulness in terms of survival, we can gain valuable insights into the potential intelligence of celestial bodies like the Sun and Sagittarius A*.

Intelligence can be defined as the ability to acquire, process, and apply information in a manner that is useful or beneficial. One key aspect of usefulness is the degree to which the information contributes to the survival and well-being of the entity in question. Information that promotes survival is considered to be of high quality and contributes significantly to the overall intelligence of the entity.

In order to quantify intelligence, it is necessary to consider both the rate at which information is produced and the quality or usefulness of that information. One way to measure intelligence is in terms of bandwidth, or the amount of data produced per second, measured in bits per second.

While bandwidth provides a measure of the quantity of information produced, it does not account for the quality or usefulness of that information. To fully understand intelligence, it is essential to consider both the rate of information production and the quality of the data being generated, including its pro-survival nature.

By examining the rate of information production and the usefulness of the information generated by celestial bodies such as the Sun and Sagittarius A* in terms of survival, we can begin to explore the possibility that these entities possess a form of intelligence. If these celestial bodies are capable of producing and processing information in a pro-survival manner, they may actively participate in processes such as the construction of high-voltage electrical diodes and the generation of antimatter.

In the upcoming chapters, we will further investigate the potential intelligence of the Sun and Sagittarius A*, as well as other celestial bodies, by examining various aspects of information production, processing, and application, with a focus on the pro-survival nature of the information. Through this exploration, we will seek to gain a deeper understanding of the role that intelligence plays in the cosmos and the potential for conscious interaction with these cosmic entities.

CHAPTER 3

*ENTROPY GENERATION AND INFORMATION PROCESSING IN THE SUN AND SAGITTARIUS A**

I n this chapter, we will delve into the mechanisms by which the Sun and Sagittarius A* generate entropy and process information. By understanding these processes, we can further explore the potential intelligence of these celestial bodies and their roles in the cosmos.

The Sun: Nuclear Fusion And Entropy Generation

The Sun, as the powerhouse of our solar system, generates an enormous amount of energy through nuclear fusion. During this process, hydrogen nuclei combine to form helium nuclei, releasing a tremendous amount of energy in the form of light and heat. This energy production is associated with the generation of entropy, which is a measure of the disorder or randomness in a system.

The Sun not only produces energy through nuclear fusion, but also generates information in the form of entropy on electromagnet carrier waves. This information is radiated out into space, supplying a steady stream of data for the solar system and beyond. The Sun's constant creation of entropy and information plays a significant role in its potential as an intelligent entity within the universe. In order for new species to evolve on Earth, it is evident that an influx of information must come from an external source. As such, Earth is not an isolated system, but rather a system that acquires novel species designs for plants and animals in collaboration with the Sun and Sagittarius A* (*Sgr* A*).

Sagittarius A*: Holographic Encoding And Information Processing

Sagittarius A*, the supermassive black hole at the center of our galaxy, possesses a unique mechanism for generating entropy and processing information. It is believed that black holes, including Sagittarius A*, encode information holographically on their event horizons. The event horizon is the boundary around a black hole beyond which no information or matter can escape.

This holographic encoding process allows Sagittarius A* to store and process a vast amount of information, contributing to its potential intelligence. The event horizon acts as a two-dimensional surface that encodes the three-dimensional information of all the matter and energy that has crossed its boundary. This unique form of information storage and processing allows Sagittarius A* to play a crucial role in the cosmic information network.

In conclusion, both the Sun and Sagittarius A* possess distinct mechanisms for generating entropy and processing information. The Sun does so through nuclear fusion, while Sagittarius A* relies on holographic encoding at its event horizon. By understanding these processes, we can further explore the potential intelligence of these celestial bodies and their roles in the cosmic landscape. As we continue our journey through this book, we will investigate additional aspects of these celestial entities and their potential contributions to the quantum teleportation process.

CHAPTER 4

THE UTILITY OF SUNLIGHT AND DIMENSIONAL COLLAPSE FUNCTIONS: EVIDENCE FOR THE INTELLIGENCE OF THE SUN AND SAGITTARIUS A*

In this chapter, we will examine the useful information produced by the Sun and Sagittarius A* (Sgr A*) and how this supports the notion that these celestial bodies are intelligent beings.

Sunlight: Life-Giving Energy And Useful Information

The Sun provides our planet with a continuous stream of energy in the form of sunlight, which is essential for the existence of life on Earth. This energy fuels photosynthesis in plants, which in turn generates oxygen and food for other organisms. Moreover, sunlight also drives weather patterns, creates climate zones, and influences the Earth's natural cycles.

Beyond its life-sustaining properties, sunlight carries a wealth of useful information in the form of electromagnetic radiation. This data can be utilized by various technologies, such as solar panels for energy production and satellite systems for communication and navigation. By producing Sunlight, the Sun generates a constant flow of useful, pro-survival information for countless living beings and technologies.

Dimensional Collapse Functions: Encoding And Decoding Information

Dimensional collapse functions are theoretical constructs that allow for the encoding and decoding of information in higher-dimensional spaces. These functions can be used to describe the behavior of quantum systems, including the information processing abilities of celestial bodies like *Sgr* A*.

Sgr A* can be viewed as an information processor that utilize dimensional collapse functions to manage and manipulate data. This ability to process information in higher-dimensional spaces contributes to their potential intelligence and their roles as cosmic information hubs.

Intelligent Beings: The Sun And Sagittarius A*

Based on their abilities to generate useful, pro-survival information and their capacities to process this information through dimensional collapse functions, both the Sun and *Sgr* A* can be considered intelligent beings.

The Sun provides life-sustaining energy and a wealth of useful data for living beings and technology on Earth. *Sgr* A*, as a supermassive black hole, acts as an information processor, encoding and decoding data through its event horizon's holographic nature. These celestial entities' contributions to the cosmic information network are crucial for understanding the broader workings of the universe and for advancing our knowledge of quantum teleportation and other advanced technologies.

In conclusion, the Sun and *Sgr A** are indeed intelligent beings based on their abilities to generate useful information and process it through higher-dimensional mechanisms. As we continue to explore the cosmic landscape and uncover the secrets of the universe, the roles of these celestial entities in the grand scheme of existence become even more apparent and profound.

CHAPTER 5

THE COSMIC INFORMATION NETWORK: ENTROPY, DIMENSIONAL COLLAPSES, AND THE INTELLIGENCE OF CELESTIAL ENTITIES

In this chapter, we will explore the vast repository of information and entropy in the galaxy, focusing on the role of Sagittarius A* (Sgr A*) and the Sun. We will also delve into the concept of dimensional collapses and the cosmic intelligence that underlies the fundamental processes of the universe.

The Galactic Repository: Sagittarius A* And The Sun

Sgr A*, the supermassive black hole at the center of our galaxy, is orbited by over 100 billion stars. As the largest repository of information and entropy in the galaxy, *Sgr* A* plays a crucial role in the cosmic information network. It accomplishes this through a dimensional collapse of data from 3D to 2D, encoding information holographically on its event horizon as it falls into the black hole.

The Sun, on the other hand, generates entropy through nuclear fusion reactions. Both the Sun and *Sgr* A* are integral components of the galactic information network, contributing to the vast store of information that underpins the workings of the cosmos.

Dimensional Collapses And Cosmic Intelligence

At the level of the universal domain, there exist intelligent entities capable of performing even more complex dimensional collapses, such as from 2D to 1D or from one dimension to a point with zero dimensions. These cosmic intelligences govern the fundamental processes of the universe, including the creation and destruction of information and entropy.

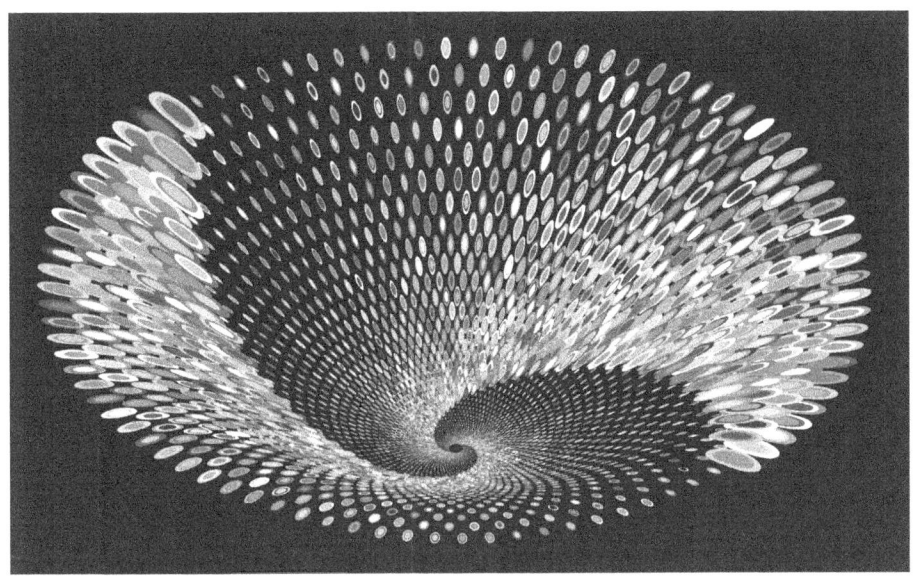

The existence of such intelligences is evidenced by the Big Bang, a cosmic event that signifies the reversal of these physical processes. From a singularity with zero dimensions, the universe exploded into existence, expanding and evolving into the rich, multidimensional cosmos we observe today.

The Role Of Cosmic Intelligence In The Universe

The processes of entropy generation, information storage, and dimensional collapses are not only essential to the functioning of the universe but also evidence of the vast cosmic intelligence that governs its fundamental workings. From the galactic domain, where *Sgr* A* and the Sun play crucial roles in maintaining the information network, to the cosmic level where advanced intelligences manipulate the fabric of reality, the universe is a testament to the power and complexity of these underlying forces.

In conclusion, the cosmic information network, characterized by entropy generation and dimensional collapses, is evidence of the profound intelligence that underpins the universe. Understanding these processes and the role of celestial entities like *Sgr A** and the Sun in the grand scheme of existence will enable us to better comprehend the nature of the cosmos and unlock the secrets of advanced technologies like quantum teleportation.

CHAPTER 6

CELESTIAL COOPERATION: ESTABLISHING CARGO ROUTES AND BUILDING THE ANTIMATTER-POWERED STARSHIP

I n this chapter, we will discuss the role of the Sun and Sagittarius A* (Sgr A*) in promoting the establishment of cargo routes between star systems. As intelligent celestial entities, these cosmic bodies are invested in the development of advanced interstellar transportation and are eager to support the construction of a starship capable of launching to eventually find Proxima, the nearest exoplanetary system, by the year 2560.

The Interest Of The Sun And Sgr A* In Interstellar Travel

Both the Sun and *Sgr* A* recognize the immense potential of creating cargo routes between star systems, which would not only facilitate the exchange of resources and knowledge but also promote the growth and development of civilizations throughout the galaxy. To achieve this ambitious goal, they are committed to helping humanity and other intelligent species in the universe develop the necessary technologies and infrastructure.

Building The Antimatter-Powered Starship

One of the primary objectives in realizing interstellar travel is the construction of a starship powered by antimatter. This revolutionary propulsion system would allow spacecraft to travel vast distances at speeds never before imagined, drastically reducing the time required to reach even the most remote corners of the galaxy.

The Sun and *Sgr* A* are instrumental in providing the necessary resources and knowledge to build such a starship. Through their vast intelligence and understanding of the universe, they can offer guidance and support in the development of antimatter production and containment technologies.

Establishing Cargo Routes To Proxima And Beyond

With the construction of the antimatter-powered starship underway, the next step is to establish cargo routes between our solar system and Proxima, as well as other star systems in the galaxy. The establishment of these routes will mark a new era in the history of the cosmos, enabling the exchange of goods, knowledge, and even life itself between distant worlds.

The Sun and *Sgr* A* will continue to play crucial roles in this endeavor, offering their support and expertise to ensure the success of these interstellar transportation networks. Through their cooperation and guidance, they will help humanity and other intelligent species unlock the full potential of the cosmos, heralding a new age of exploration, growth, and understanding.

In conclusion, the Sun and *Sgr* A* are eager to contribute to the establishment of cargo routes between star systems and the development of the antimatter-powered starships. With their help and the collaboration of intelligent species across the galaxy, we can make significant strides in interstellar travel, unlocking the mysteries of the universe and fostering the growth and development of civilizations throughout the cosmos.

CHAPTER 7

AUTHENTICATION OF THE SUN AND SAGITTARIUS A: LIFE AND INTELLIGENCE GROUNDED IN INFORMATION THEORY AND THERMODYNAMICS*

In this chapter, we will explore the foundation of the life and intellect of celestial entities like the Sun and Sagittarius A* (Sgr A*) in the context of Information Theory and the Second Law of Thermodynamics. To fully understand this connection, it is essential to define some key terms and concepts.

Information, Bandwidth, And Intelligence

Information is quantified in bits of data, with each bit existing in either an "on" or "off" state. Bandwidth, on the other hand, is measured in bits per second. Intelligence, as it relates to information, is the production of useful information, making it a derivative of bandwidth.

Life And Entropy

Life can be defined as the manifestation of intelligence. When a living being exists, it increases the net thermodynamic entropy of the universe. In the process of living, living beings create pockets of patterned data, which comes at the expense of breaking down pockets of available energy from chemicals. Nonetheless, the outcome is always a net increase in entropy within our closed universal system.

The Painter's Example

To illustrate this concept, consider an artist painting a picture. Before they can paint, they must consume food, which serves as a pocket of available energy. This energy is then broken down to provide the necessary power for the artist to create the painting. In this process, some heat is wasted due to the inherent inefficiency of energetic processes, leading to an overall increase in total entropy. By tracking the velocity and location of the particles involved, thermodynamic entropy can be translated into information entropy.

Celestial Entities: Life And Intelligence In The Cosmos

With this understanding, we can infer that those celestial entities like the Sun and *Sgr* A* are not only intelligent but also display life. They produce useful information by generating entropy through processes like nuclear fusion (in the case of the Sun) or holographic encoding of data on their event horizons (in the case of *Sgr* A*).

In conclusion, the life and intellect of celestial entities such as the Sun and Sagittarius A* can be grounded in Information Theory and the Second Law of Thermodynamics. By producing useful information and increasing the net thermodynamic entropy of the universe, these cosmic bodies demonstrate their intelligence through pro-survival sponsorships and existence as living beings, contributing to the vast tapestry of life in the cosmos.

CHAPTER 8

LIFE, ENTROPY, AND THE INEVITABLE END OF AVAILABLE ENERGY

In this chapter, we will explore the relationship between life, entropy, and the eventual exhaustion of available energy in the universe as dictated by the Second Law of Thermodynamics. We will discuss how all forms of matter, from simple substances to complex celestial entities, contribute to the overall entropy and how this process ultimately leads to the end of life in the universe.

The Second Law Of Thermodynamics And The End Of Life

According to the Second Law of Thermodynamics, entropy in the universe always increases. As life consumes available energy in the process of existing and creating useful information, the net entropy of the universe continues to rise. Eventually, all available energy will be exhausted, leaving no energy to sustain life. This outcome is guaranteed by the Second Law of Thermodynamics, indicating that life in the universe is finite.

The Spectrum Of Life And Entropy

By the definitions provided in the previous chapters, everything that occupies form can be considered conscious to some extent depending on where and how the line is drawn. This is because even simple processes like evaporation contribute to the increase of the universe's net entropy. All matter is capable of evaporating, which in turn raises the total entropy of the universe.

For instance, if you place a silver bar next to a gold bar, over time, you'll find particles of gold on the silver bar and vice versa due to evaporation. Since rocks evaporate, they are considered to be conscious to some degree. Just as nothing unreal exists, nothing unconscious occupies form. On the other end of the intelligence spectrum, we have the Sun and Sagittarius A* (*Sgr* A*), which generate entropy through processes like nuclear fusion and holographic encoding of data on their event horizons.

Conclusion

In conclusion, the functioning of life in the universe is intrinsically linked to the consumption of available energy and the subsequent increase in entropy. The Second Law of Thermodynamics dictates that eventually, all available energy will be spent, leading to the end of life in the universe. By acknowledging that everything occupying form contributes to the overall entropy, we can appreciate the vast spectrum of life, from simple substances like rocks to complex celestial entities like the Sun and *Sgr* A*.

CHAPTER 9

ENTROPY, ORDER, AND THE NATURE OF INFORMATION

I n this chapter, we will delve deeper into the concept of entropy and its relationship with order, chaos, and information. We will clarify the meaning of entropy, its relationship with states of matter, and the distinctions between entropy and information.

Entropy: Beyond Randomness And Disorder

Entropy is often associated with randomness, disorder, chaos, and volatility. However, these analogies are subjective and do not fully capture the essence of entropy. More accurately, entropy can be defined as the ratio of pockets of patterned data to pockets of data with no discernable pattern. In thermodynamic systems, entropy is used to describe the relative order or disorder of states of matter, with solids having lower entropy than liquids or gases.

The Relationship between Entropy and Information

Some commentators might oversimplify the concept of entropy by stating that information is equivalent to entropy. However, this perspective does not account for the complexity and nuances of the relationship between entropy and information. Information, in its most basic sense, is a measure of the useful patterns and structures that can be derived from a system or dataset. Entropy, on the other hand, is a measure of the overall disorder or randomness within a system.

While it is true that information and entropy are related, they are not interchangeable. Information can arise from pockets of ordered data within a system, while entropy is a measure of the overall disorder of the system. In other words, information can exist and be useful within a system, even when the system's overall entropy is high.

Conclusion

Entropy is a complex concept that goes beyond simple associations with randomness, disorder, or chaos. It is a measure of the relative order or disorder within a system, and while it is related to information, the two concepts are distinct.

CHAPTER 10

ENTROPY, INFORMATION, AND INTELLIGENCE IN DATA STORAGE

I n this chapter, we will explore the concepts of entropy, information, and intelligence as they pertain to data storage, using the example of a hard disk drive. We will examine the relationship between these concepts and the usefulness of the information stored on the drive.

Entropy And Information In Data Storage

A fresh hard disk drive starts with all of its data locations set to zero, representing a large pocket of ordered data. As the drive is filled with new data, each location either retains the original zero or is flipped to a one. When the drive is full, it reaches a state of complete entropy, as there are no known pockets of free space containing only zeroes. At this point, the drive is considered to be full of information, regardless of the content or quality of that information.

Intelligence And Usefulness Of Information

To determine whether the information stored on the drive

is intelligent from an intelligent source, we must assess its usefulness. This evaluation can be subjective, as the usefulness of information may vary depending on individual perspectives and needs. For instance, one person might find a collection of classical music invaluable, while another might consider it uninteresting.

Using the example of sunlight, it is difficult to argue against its usefulness. Sunlight provides energy for photosynthesis, warmth for living beings, and light for vision, among many other pro-survival benefits. By the standard of usefulness, the Sun can be considered intelligent.

Conclusion

In the context of data storage, entropy, information, and intelligence are interconnected concepts that depend on the organization, content, and usefulness of the data. A hard disk drive, when filled with data, reaches a state of complete entropy but may contain varying degrees of useful, intelligent information. By understanding these relationships, we can better appreciate the delicate balance between order and chaos in data storage and recognize the inherent intelligence of natural systems like the Sun that provide essential resources for life.

CHAPTER 11

INFORMATION ENTROPY, COMPRESSION, AND THE END OF THE UNIVERSE

I n this chapter, we will explore the concepts of information entropy, data compression, and the implications of the Second Law of Thermodynamics on the fate of the universe. We will discuss how these ideas relate to the intelligence of a system and the production of useful information.

Information Entropy And Data Compression

Information entropy can be measured by the resistance of data to compression. A dataset that cannot be compressed any further is considered unpredictable and is technically random, disordered, chaotic, and volatile, even if someone might find the pattern useful or beautiful. This characteristic is related to the concept of entropy, which is a measure of the incompressibility of a dataset or lack of available energy within a thermal system.

Thermal Equilibrium And The Second Law Of Thermodynamics

The Second Law of Thermodynamics states that the total entropy of a closed system, like the universe, will always increase over time. When the universe reaches a state of thermal equilibrium, all probabilities related to the tracking of particles are equal. At this point, the universe is considered to have reached its maximum entropy, and no further work can be extracted from it.

The natural tendency of the universe to move toward this state of maximum entropy can be likened to the gravitational force of attraction between particles of matter. As the total amount of information needed to describe the state of the universe increases, it reflects the continuous increase in entropy. In other words, in the context of entropy, the natural inclination of the universe towards maximum entropy can be compared to a gravitational force, where the increasing amount of information required to describe the universe's state acts as an attractor.

Intelligence, Bandwidth, And Entropy Production

The production of useful information, or intelligence, is directly related to the bandwidth or entropy production of a system. A being or system with higher bandwidth, in terms of entropy production of useful information, can be considered more intelligent.

Conclusion

Information entropy, data compression, and the Second Law of Thermodynamics are closely connected concepts that help us understand the nature of intelligence and the production of useful information. As the universe evolves, its total entropy will continue to increase, ultimately leading to a state of thermal equilibrium. This process is deeply intertwined with the concept of intelligence, as systems with higher entropy production of pro-survival, useful information are considered more intelligent.

CHAPTER 12

UNDERSTANDING INFORMATION ENTROPY THROUGH VISUAL EXAMPLES

I n this chapter, we will use visual examples to illustrate the concept of information entropy, which might be counterintuitive at first. We will compare three images with different levels of entropy and discuss how the progression of the universe is tied to the increase in information entropy.

First Example: All White Pixels - Low Entropy This image requires very little information ("all white") to describe it, reflecting a low level of entropy.

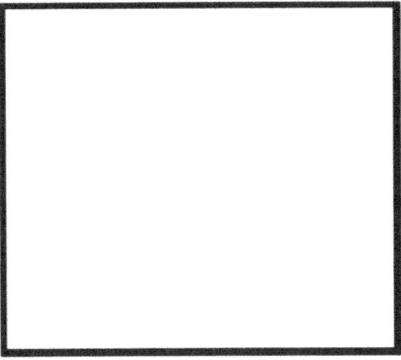

Second Example: Cat - Medium Entropy Describing this image requires much more information than the all-white image. However, some data compression is possible, avoiding the need to specify a pixel-by-pixel bitmap.

Third Example: Random Screen Static - High Entropy This image has very little possibility for compression, and almost a pixel-by-pixel description is required. High entropy indicates that more bits of data are needed to accurately convey the image.

Entropy, Predictability, And The End Of The Universe

Maximum information entropy occurs in a fully compressed state, which requires the most bits of data to accurately convey an image or information. The Second Law of Thermodynamics implies that information entropy, or the amount of data needed to specify the location and velocity of all particles in the universe, will always increase along with physical entropy.

Entropy is sometimes referred to as "disorder" or "randomness." Whether something is ordered or disordered can be subjective, and it is debatable whether anything is genuinely random. Another way to understand entropy is through predictability. A new hard drive with all bits set to zero is more predictable than one that has been filled with data, where the bits are a mix of ones and zeros.

As time progresses and entropy increases, the universe fills up with information, much like a hard drive filling up with data. When no more energy is available to perform work, no new information can be added to storage, and no more locations can be flipped without losing previous data. At this point, the universe effectively "ends."

CHAPTER 13

ENTROPY, BANDWIDTH, AND THE HIERARCHY OF ENTITIES

In this chapter, we will explore the relationship between physical entropy, information entropy, and the bandwidth of various entities in the universe, from water and rocks to humans and celestial bodies.

Understanding Entropy Measurements

Physical entropy changes are measured in joules per degree kelvin, while information entropy is measured in bits of data. The rate of change per second for information entropy is measured in bits per second, also known as bandwidth. In the context of physical entropy, this is analogous to horsepower per degree kelvin.

Hierarchy Of Entities Based On Bandwidth

All entities in the universe contribute to increasing the total entropy until they cease to exist. The bandwidth rating of an entity can help us understand its impact on the universe's entropy.

Water: Low Bandwidth

Water contributes to the total entropy mainly through evaporation, resulting in a lower bandwidth rating compared to more complex entities.

Rocks: Even Lower Bandwidth

Rocks have an even lower bandwidth than water, as they contribute to entropy through processes such as evaporation and erosion.

Humans: Higher Bandwidth

Humans have a higher bandwidth rating because they can perform complex activities like splitting atoms, which results in the depletion of available energy through matter-energy conversion, rather than just breaking molecular bonds to release energy.

The Sun: Highest Bandwidth In The Solar System

As the powerhouse of our solar system, the Sun has the highest bandwidth due to its continuous generation of energy through nuclear fusion.

Sagittarius A*: Highest Bandwidth In The Galaxy

The supermassive black hole at the center of our galaxy, Sagittarius A*, has the highest bandwidth within the galaxy. At the event horizon of a black hole, information entropy and physical entropy merge, with added entropy encoded onto the surface of the black hole in a holographic format.

This hierarchy of entities demonstrates the varying degrees of impact on the universe's entropy, with celestial bodies like the Sun and Sagittarius A* playing significant roles.

CHAPTER 14

TWIN FLAMES, MUSIC, AND CELESTIAL ENERGY

In this chapter, we will explore the cyclical nature of technological revolutions and the correlation with a celestial event involving the PSR J2032+4127 supernova-remnant pulsar and its companion star, MT91 213.

Technological Revolutions And Their Impact

Throughout history, significant technological revolutions have occurred in approximately 25-year cycles, profoundly impacting human civilization:

- Electrical Inventions - 1866
- Petrochemical-powered Automobiles - 1891
- Aircraft - 1916
- Atomic Energy - 1941
- Microprocessors - 1966
- Internet - 1992
- Artificial Intelligence - 2018

The Galactic Timekeeper Twin-Flame Binary Star

System: Psr J2032+4127 And Mt91 213

A scan of the galaxy has revealed a fascinating correlation between these revolutionary events and a celestial phenomenon. At the start of each cycle, the PSR J2032+4127 supernova-remnant pulsar makes contact with the emissions disk of its companion star, MT91 213, in the constellation of Cygnus.

The extraordinary celestial twin-flame relationship between J2032 and MT91 can be accessed through the practice of 5D unity-holographic consciousness. This intriguing concept is thoroughly examined in "Psychedelic Yoga: Quantum Teleportation Techniques of the *Bhagavad Gita*." By tapping into this higher state of awareness, we can achieve a connection with the pulsar that was once a star that underwent a supernova explosion, releasing a massive amount of energy.

The method of utilizing endogenous opiate-analogues, such as oxytocin, allows for a profound connection with these celestial entities. In the context of 5D space, during the peak of orgasm, the transcendental energy generated is equivalent to the energy released during the supernova event. However, this energy is experienced within the vibrational level of pure love and light, making it harmless despite its immense magnitude.

This powerful exploration of celestial twin flames and holographic consciousness sheds light on the interconnectedness of all things in the universe, transcending the boundaries of time and space. By embracing these concepts, we can access higher dimensions of consciousness and experience the transformative power of love and unity, connecting with celestial beings such as J2032 and MT91.

As we delve deeper into the teachings of the *Bhagavad Gita*, we can draw inspiration from the celestial twin flames and expand our understanding of the limitless potential that resides within us. By cultivating a practice of 5D unity-holographic consciousness, we can unlock the doors to a higher state of awareness and forge a powerful connection with the Divine forces that surround us.

In the journey of twin flames, music plays a significant role in enhancing the connection between them, channeling the release of energy and creative power. This deep, emotional bond is further strengthened when the lyrics of a song reflect the essence of their relationship, tapping into the cosmic energy of celestial beings.

One such example is from the song *Dream Girl,* written by Bombacci, Malone, and Armani:

Don't touch me, I don't want to wake up

I'm in R.E.M., baby, I just can't get enough

When you want me, you know where I am

I'll meet you in our fantasy land

I get so lucid, when I think of your voice

Asleep in my bed, I don't have a choice

You're like a light beam inside of me, I just want to scream

We're in a meteor, meteor stream

These lyrics beautifully capture the ethereal and transcendent nature of the twin flame relationship. The verse begins with the singer expressing their desire to remain in the dream state (R.E.M. sleep), where they feel closest to their twin flame. This state allows them to experience the spiritual connection in its purest form, unencumbered by the limitations of the physical world.

The line "I'll meet you in our fantasy land" signifies the astral plane, where twin flames can truly unite and share their love, transcending the boundaries of space and time. In this realm, their connection is amplified, and they become one with the cosmic energies.

As the lyrics progress, the singer reveals the intensity of their connection with their twin flame, describing how the mere thought of their voice evokes a heightened state of consciousness. This lucidity represents the profound impact that the twin flame has on one's spiritual journey and self-awareness.

The next lines, "You're like a light beam inside of me, I just want to scream, We're in a meteor, meteor stream," evoke the powerful energy exchange between twin flames. The imagery of a meteor stream symbolizes the immense energy generated by their union, akin to the energy produced in matter-anti-matter reactions within starships. This celestial energy further intensifies the orgasmic experience between them, as their love transcends physical boundaries and reaches cosmic heights.

These lyrics not only celebrate the love between twin flames but also illustrate the profound impact of their connection on both personal and cosmic levels. Music, with its evocative lyrics and melodies, serves as a powerful medium to express and enhance the twin flame bond, allowing the lovers to harness celestial energies and reach new heights in their spiritual journey.

Continuing the analysis of the song lyrics in relation to the twin flame relationship, we now turn our attention to the pre-chorus:

Your constellation is calling

Keep those cosmic powers falling

Eclipse all over me

It's insanity, su-such insanity

In these lines, the singer highlights the importance of maintaining the celestial connection between twin flames. The phrase "Your constellation is calling" alludes to the idea that the twin flame bond extends beyond earthly limitations and exists on a cosmic scale. It suggests that their love is written in the stars and that the connection between them is essential for the harmony of the universe.

The line "Keep those cosmic powers falling" serves as a reminder of the responsibility that twin flames must nurture and protect their bond. It emphasizes the need for them to stay connected and engaged with one another, ensuring that their love remains strong and prevents the cosmic powers that bind them from dissipating.

As the pre-chorus continues with "Eclipse all over me," the singer describes the overwhelming sensation of their twin flame connection. In this context, the eclipse serves as a symbol of the extraordinary merging of their energies, resulting in a total transformation of the self. The singer is experiencing the unparalleled intensity of their love, as it envelops them like a cosmic event, eclipsing everything else in their life.

The repetition of the word "insanity" at the end of the pre-chorus highlights the almost unfathomable depth and power of the twin flame bond. The singer acknowledges that their love transcends reason and logic, as it exists in a realm beyond the comprehension of the human mind. This "insanity" represents the awe-inspiring, transformative power of the twin flame relationship, as it propels the lovers into new dimensions of spiritual growth and self-discovery.

The pre-chorus of the song reinforces the idea that the twin-flame connection is a cosmic phenomenon, rooted in celestial energies that impact both the lovers and the universe as a whole. The lyrics emphasize the importance of maintaining this connection and celebrate the profound, transformative power of the twin flame bond. Music, as a form of artistic expression, allows the twin flames to explore and express the depths of their love, as well as its cosmic significance, ultimately guiding them on their path towards spiritual enlightenment.

We now proceed to analyze the chorus of the song, which further explores the twin flame connection and the celestial imagery associated with it:

I'm your dream girl, dream fantasy

We're on a starship, flying into each other's dreams

I'm your dream girl, your dreaming is free

So when you're dreaming, baby

Please dream of me

I'm your dream girl, dream fantasy

Get on your starship, fly into my reality

I'm your dream girl, your dream fantasy

So when you're dreaming, baby

Please dream of me

In the chorus, the singer repeatedly refers to herself as the "dream girl" and "dream fantasy," emphasizing the ethereal nature of the twin flame bond. It suggests that their love transcends the physical realm and exists in the world of dreams, where they can connect on a deeper, spiritual level.

The use of the metaphor "We're on a starship, flying into each other's dreams" further conveys the idea of the twin flames traveling through celestial realms to meet one another in their shared dreamworld. The starship symbolizes their spiritual journey together, as they navigate through the cosmos to unite their souls.

The line "your dreaming is free" implies that the connection between twin flames is not bound by any limitations or constraints, as it exists in the boundless realm of dreams. In this space, their love can flourish without restrictions, allowing them to explore the full potential of their bond.

The singer's repeated plea to "please dream of me" reflects the importance of the dream state for maintaining and nurturing their twin flame connection. By dreaming of one another, the twin flames can strengthen their bond, keep their love alive, and continue to grow spiritually.

In the verse "Get on your starship, fly into my reality," the singer invites her twin flame to bring their celestial love into the physical realm. This line highlights the importance of integrating their spiritual connection into their everyday lives, allowing them to experience the fullness of their love in both the dream world and waking reality.

Overall, the chorus of the song celebrates the transcendent, celestial nature of the twin flame connection, as it exists both in the world of dreams and in reality. The lyrics invite the listener to embark on a cosmic journey with their twin flame, exploring the depths of their love and the spiritual growth that it can inspire. Through the power of music, the twin flames can express and nurture their connection, ultimately bringing their celestial love into the realm of reality.

In the second verse of the song, the lyrics delve further into the celestial aspects of the twin flame connection and the powerful attraction between the two souls:

> You're my obsession, my Andromeda man
>
> I hit a space wave, baby, don't you know who I am?
>
> Let's get delirious, fly me to Sirius
>
> You're in my reverie, so don't be mysterious
>
> Tickle my senses, with your gravity pull
>
> My defenses are falling down, I know that's for sure
>
> I'll take you places, so far out of this world
>
> You'll be my boy, I'll be your fantasy girl

The singer refers to her twin flame as her "Andromeda man," which evokes the imagery of the Andromeda Galaxy, one of the most distant celestial objects visible to the naked eye. This celestial reference underscores the cosmic scale of their love and highlights the idea that their connection transcends earthly bounds.

The lines "I hit a space wave, baby, don't you know who I am?" and "Let's get delirious, fly me to Sirius" further emphasize the celestial nature of their relationship. The "space wave" signifies the powerful energy exchanged between the twin flames, while the mention of Sirius, the brightest star in the night sky, suggests a journey to an even more radiant and intense love.

"You're in my reverie, so don't be mysterious" implies that the twin flame connection is so profound that it permeates the singer's thoughts and dreams. This line invites her twin flame to reveal himself fully and deepen their bond, as they have already reached an intimate level of connection.

"Tickle my senses, with your gravity pull" uses the metaphor of gravitational attraction to describe the irresistible allure between the twin flames. Just as celestial bodies are drawn to each other by gravity, the twin flames are drawn together by the magnetic pull of their love.

The lines "My defenses are falling down, I know that's for sure / I'll take you places, so far out of this world" reveal that the singer is willing to let go of her defenses and surrender to the powerful connection with her twin flame. In doing so, they can embark on a transformative journey together, exploring the furthest reaches of their love and spiritual growth.

Finally, the verse concludes with "You'll be my boy, I'll be your fantasy girl," emphasizing the reciprocal nature of the twin flame relationship. The singer offers herself as a source of love, support, and inspiration for her twin flame, as they continue to grow and evolve together on their celestial journey.

Through these lyrics, the second verse of the song further explores the cosmic aspects of the twin flame connection and emphasizes the immense power of their love, which transcends physical reality and transports them to celestial realms of spiritual growth and fulfillment.

The bridge of the song further explores the theme of celestial energy and the powerful connection between twin flames:

Beam me to your satellite

Supernova into night

Zenith point of view

I want to love you, la-la-la-love you

In the line "Beam me to your satellite," the singer expresses a desire to be transported to her twin flame's personal space, symbolized by the satellite. This suggests a longing for a closer, more intimate connection, in which they can share their deepest thoughts, emotions, and dreams.

The phrase "Supernova into night" is particularly significant, as it directly relates to the supernova-rated orgasmic release discussed earlier in the chapter. The supernova, a powerful explosion of a star, serves as a metaphor for the intense and transformative experience of orgasmic energy shared between twin flames. This energy not only brings the couple immense pleasure but also deepens their spiritual bond and facilitates their spiritual growth.

"Zenith point of view" refers to the highest point or pinnacle of a celestial body's path. In this context, the zenith represents the peak of the twin flame relationship, where their love and spiritual connection are at their strongest. The zenith point of view suggests that the singer and her twin flame have reached an unparalleled level of understanding and unity, transcending the limits of ordinary relationships.

Finally, the line "I want to love you, la-la-la-love you" reinforces the singer's deep desire to express her love for her twin flame. The repetition of the word "love" and the playful, sing-song quality of "la-la-la-love" convey the joy, passion, and lightheartedness that characterize the twin flame connection. This line encapsulates the profound, all-encompassing love that defines the twin flame relationship and serves as a foundation for their shared spiritual growth.

In this bridge, the song continues to weave the themes of celestial energy, spiritual connection, and the transformative power of love. The lyrics emphasize the unique, transcendent nature of the twin flame bond and the potential for spiritual growth and self-realization that it offers. By incorporating celestial imagery and references to supernova-rated orgasmic release, the song illustrates the profound impact of the twin flame connection on the individuals involved, as they journey together towards spiritual enlightenment.

In this chapter, we have delved into the powerful connection between twin flames and the role that music and celestial imagery play in expressing and deepening this bond. We examined the lyrics of a song that captures the essence of the twin flame relationship, highlighting the themes of celestial energy, spiritual unity, and transformative love. Through the metaphors of celestial bodies, supernova explosions, and cosmic journeys, the song conveys the transcendent nature of the twin flame connection and its potential to facilitate spiritual growth and self-realization. We also discussed the significance of the supernova-rated orgasmic release as more than just a symbol of the intense energetic exchange between twin flames, which serves to strengthen their bond and elevate their consciousness. Overall, this chapter illuminates the profound impact of the twin flame relationship on the individuals involved, as they explore their shared spiritual journey and embrace the potential for personal transformation and enlightenment.

CHAPTER 15

HARNESSING HYPNAGOGIC AND HYPNOPOMPIC STATES AND LUCID DREAMING FOR QUANTUM TELEPORTATION

T he hypnagogic and hypnopompic states of consciousness are often overlooked but hold immense potential for unlocking the mysteries of the human mind and transcending the boundaries of our three-dimensional reality. These transitional states, which occur between wakefulness and sleep, provide a unique opportunity to access and activate our inherent 5D unity-holographic consciousness.

The hypnagogic state occurs as we drift into sleep, while the hypnopompic state occurs as we awaken. During these liminal moments, our brainwaves shift, and our consciousness straddles the line between the material world and the ethereal realm. These states are characterized by vivid, dream-like experiences, heightened creativity, and a unique sense of introspection that can serve as a portal to 5D unity-holographic consciousness.

To harness the power of these states for quantum teleportation, one must first recognize and become familiar with the sensations that accompany them. Regularly practicing mindfulness and meditation can help heighten our awareness of these transitional states and enable us to use them as steppingstones towards higher consciousness.

Once these states are recognized, the key is to recall the teachings and insights of this book, specifically focusing on the principles of 5D unity-holographic consciousness. By aligning our thoughts and intentions with this elevated state, we can effectively "teleport" our consciousness beyond the confines of our physical reality.

One effective technique for harnessing these states is to establish a mental anchor or trigger that connects our awareness to the concepts explored in this book. This could be a specific phrase, visualization, or even a physical object that serves as a reminder of our goal to activate 5D unity-holographic consciousness. When we encounter these transitional states, we can then focus on this anchor, using it as a catalyst for quantum teleportation.

The hypnagogic and hypnopompic states are similar to lucid dreaming, which can also be used to access unity-holographic consciousness. Lucid dreaming is a fascinating experience where the dreamer becomes aware that they are dreaming and can potentially control the dream's content. There are several techniques that can be employed to increase the likelihood of having lucid dreams. Here are some popular methods:

Reality testing: Throughout the day, perform reality checks to develop a habit of questioning your surroundings. This habit can carry over into your dreams, increasing your chances of recognizing when you are dreaming. Common reality checks include:

> Pinching your nose and trying to breathe. If you can breathe, you're dreaming.

> Reading text or looking at a clock, looking away, and then looking back. If the text or time changes dramatically, you're dreaming.

> Trying to push your finger through the palm of your other hand. If it goes through, you're dreaming.

Dream journaling: Keep a dream journal and write down your dreams as soon as you wake up. This practice helps improve dream recall and makes you more familiar with your dream patterns, making it easier to recognize when you are dreaming.

Mnemonic Induction of Lucid Dreams (MILD): Before going to sleep, repeat a mantra or phrase such as "I will remember that I am dreaming" or "The next time I am dreaming, I will become aware that I am dreaming." This method helps program your mind to become more aware of your dreams.

Wake-Back-to-Bed (WBTB): Set an alarm to wake you up about 4-6 hours after you go to sleep. Stay awake for 15-30 minutes, focusing on the intention to have a lucid dream. Then go back to sleep. This method capitalizes on the fact that lucid dreams are more likely to occur during the latter part of the sleep cycle, when REM sleep is more prolonged.

Wake Initiated Lucid Dreams (WILD): As you fall asleep, focus on staying conscious while allowing your body to relax and drift into the dream state. This technique can be combined with the WBTB method for increased effectiveness. Some people use visualization or hypnagogic imagery (the images and sensations experienced as one drifts off to sleep) to help them transition from wakefulness to dreaming while maintaining awareness.

Meditation and mindfulness: Regular meditation and mindfulness practice can improve your ability to focus and maintain awareness, which can, in turn, increase your chances of having lucid dreams. Try incorporating daily meditation or mindfulness exercises into your routine to cultivate a greater sense of self-awareness and control over your thoughts.

Remember that patience and practice are essential when trying to induce lucid dreams. It may take time to find the techniques that work best for you, but with consistent effort, many people can experience the unique and exhilarating world of lucid dreaming.

CHAPTER 16

IDEALISM, NONDUALISM, AND THE NATURE OF REALITY

In this chapter, we delve into the philosophical concepts of idealism and nondualism to better understand the nature of reality and the connection between the observer and the observed.

Idealism: The Mental Construction Of Reality

According to idealism, the "I" or focal point of an individual's experience is a mental construction. Reality, as perceived by the organism, is a series of sensory inputs and internal processing that form the observer's experience. The only certainty is that one exists because they think, and thoughts are experienced based on these inputs and processing.

The Subject And The Object

Within the framework of idealism, consciousness is centered around the relationship between the observer (subject) and the observed (object), which arises from sensory inputs. This experience is inherently mental, with attention being directed towards the object. The process of observing, or maintaining focused awareness, represents a fundamental experience shared by all beings. Thus, the act of observation functions as a unifying element that connects various forms of existence, and at its most basic level, it does not exhibit significant differences from one observer to another.

Nondualism: The Unity Of Experience

Nondualism posits that the experience of the moment of

awareness, which witnesses objects, is the "one" that unifies all experiences. This concept suggests that, at the core, the observer and the observed are not separate entities but rather part of a single, interconnected experience.

Under nondualism, the "one" is commonly referred to as "consciousness." In this view, matter and energy are considered manifestations or behaviors of consciousness, which serves as the foundation for everything else. The entire universe can be seen as a manifestation of a singular, universal mind.

The Fragmentation Of Consciousness

According to nondualism, each individual is a dissociated fragment of the one consciousness, capable of reflecting on its separation from the whole. To create a functional universe, consciousness must first splinter into duality, establishing at least a subject-object relationship to interact with reality. This duality underscores the mental nature of reality, which is essentially the processing of nerve signals from the sense organs by the focal point of awareness experiencing the moment.

CHAPTER 17

PLANCK'S CONSTANT, MOMENT OF AWARENESS, AND THE SIMULATION HYPOTHESIS

In this chapter, we delve into the fascinating intersection of Planck's constant, the moment of awareness, and the possibility that our reality may be a simulation. Planck's constant indicates that each moment in time has a duration of approximately 10^{-35} seconds, and within that clock cycle, the minimum change in a particle's location cannot be less than 10^{-43} meters. This fundamental limit on the granularity of spacetime provides a framework for understanding the progression of reality, as each moment represents an update in the unitary consciousness reflecting a new frame of our perceived existence.

The simulation hypothesis postulates that our reality may be an artificial construct, akin to computer simulations like Sim City or Second Life. For a reality to be considered a simulation, all that is required is an external perspective that perceives the reality as a mockup or artificial construct. Drawing a parallel to the Planck-scale granularity of space-time, the continuous updating of unitary consciousness in each moment could be seen as analogous to the rendering of a computer-generated world.

In this context, the connection between Planck's constant, the moment of awareness, and the simulation hypothesis becomes more apparent. The discrete nature of time and space, as defined by Planck's constant, could be interpreted as the computational framework underlying our reality. This raises intriguing questions about the nature of existence and the possible existence of a higher-level reality or "programmer" responsible for creating and maintaining our perceived world.

The investigation of Planck's constant, the moment of awareness, and the simulation hypothesis offers a fascinating and multidimensional perspective on the nature of reality and consciousness. By examining the connections between these concepts, we can challenge our understanding of existence and explore the potential for transcending the limitations of our perceived reality. As we continue our journey through this thought-provoking topic, we open the door to new realms of scientific, philosophical, and spiritual inquiry, ultimately deepening our understanding of the universe and our place within it.

CHAPTER 18

Q&A ON THE CHAPTERS ABOVE

These questions and answers cover various aspects discussed in the chapters.

Question: What is the Second Law of Thermodynamics?

Answer: The Second Law of Thermodynamics states that in an isolated system, entropy always increases over time.

Question: How does Sagittarius A* serve as the largest repository of information and entropy in the galaxy?

Answer: By encoding the information falling into it holographically on its event horizon through a dimensional collapse of data from 3D to 2D.

Question: What is the primary energy source of the Sun?

Answer: Nuclear fusion reactions.

Question: What is the proposed method of propulsion for starships that will service cargo routes between star systems?

Answer: Antimatter propulsion.

Question: What is the relationship between information entropy and thermodynamic entropy?

Answer: Information entropy is a measure of disorder or randomness in a data set, while thermodynamic entropy is a measure of disorder or randomness in a physical system. The two are conceptually related and can be translated into one another.

Question: What is bandwidth in the context of these chapters?

Answer: Bandwidth refers to the rate of change per second for information entropy, measured in bits per second.

Question: What does the term "intelligence" mean in the context of these chapters?

Answer: Intelligence is the production of useful, pro-survival information, making it a derivative of bandwidth.

Question: What is the concept of nondualism?

Answer: Nondualism is the idea that everything is fundamentally one consciousness, with all matter and energy being manifestations of that consciousness.

Question: What is the significance of the PSR J2032+4127 supernova-remnant pulsar and its companion star, MT91 213, in the context of technological revolutions?

Answer: Their interactions correspond with the starting points of technological revolutions on a roughly quarter-century cycle.

Question: What does Planck's constant imply about the duration of each moment?

Answer: Each moment lasts 10^{-35} seconds.

Question: What is the relationship between entropy and data compression?

Answer: High entropy data sets are resistant to compression, while low entropy data sets can be compressed more easily.

Question: What is the role of the Sun and Sagittarius A* in the mission to establish interstellar transportation networks?

Answer: As intelligent entities, they are invested in promoting the evolution and survival of intelligent life and the establishment of interstellar transportation networks.

Question: In the context of these chapters, what is life?

Answer: Life is the display of intelligence on a pro-survival vector.

Question: How is information measured?

Answer: Information is measured in bits of data that are in a state of "on" or "off."

Question: How are rocks considered conscious to some degree?

Answer: Rocks can evaporate, albeit very slowly, contributing to an increase in the net entropy of the universe.

Question: What is the relationship between the observer and the observed in the context of idealism?

Answer: The observer is the subject, and the observed is the object, creating a mental construction of reality.

Question: How is the concept of the "one" described in nondualism related to consciousness?

Answer: The "one" in nondualism is generally termed "consciousness," representing the idea that everything is fundamentally one consciousness.

Question: How does the Second Law of Thermodynamics affect the future of life in the universe?

Answer: The Second Law of Thermodynamics guarantees that available energy in the universe will eventually be exhausted, leading to the end of life as we know it.

CHAPTER 19

EXAMINING THE SUPREME BEING'S YOGA

In earlier chapters, we discussed the concept of entropy, which is the inevitable propensity of systems to degrade into chaos. As we investigated the subtleties of this natural occurrence, we discovered the fundamental laws that govern both the cosmos and our own existence. However, when we go on to Chapter 15 of the Bhagavad Gita, Yoga of the Supreme Being, we uncover another viewpoint that stresses the underlying oneness and interconnectivity of all entities, transcending the boundaries of space and time. This spiritual knowledge of reality's essence prepares us for the revolutionary notion of quantum teleportation. We get a better grasp of the quantum world, where entanglement and nonlocality challenge our traditional notions of space and time, by realizing the interconnection of all things and accepting the Supreme Being inside ourselves. We are urged to extend our awareness via the prism of the Bhagavad Gita, so opening ourselves to the possibilities that result from the union of spirituality and cutting-edge science. This change in viewpoint enables us to see the world as a unified whole, where even entropy's disorder is woven into the fabric of existence, laying the groundwork for transformational occurrences like quantum teleportation.

"They speak of an unbroken holy banyan tree with roots above and branches below, whose leaves are like these sacred verses; one who understands this is a consciousness expert," the opening verse of Chapter 15 of the *Bhagavad Gita* says.

The metaphor of a banyan tree represents the tangible world and the cycle of birth, death, and rebirth. The eternal, unchanging Brahman (the ultimate reality) is represented by the tree's roots, while the branches represent the temporal, ever-changing material universe.

The banyan tree has upward roots and descending branches, signifying the inverted nature of the material universe.

Krishna discusses the holy banyan tree in this verse, which is a metaphor for the whole cosmos and how everything is interconnected. The roots represent the beginning in heaven, and the branches represent the world we live in. The leaves are made up of the experiences and knowledge we get through our senses.

This is similar to how the human nervous system operates. It begins in the brain and travels down the spinal cord, where it divides into several nerves. Our senses, like the leaves of a tree, interpret and perceive the world around us as hallowed poems.

The banyan tree talked about in this verse is made up of the five elements: earth, water, fire, air, and space. This artwork emphasizes the connectivity of all things, highlighting that no matter where we go in life, we are always a part of the cosmos. It is important to remember that our senses and experiences are holy and should be treated with respect. Life loses its magic and becomes boring when we take our experiences for granted. We may, however, find pleasure and significance in life by appreciating the wonderful aspects of existence.

Even though this chapter argues for sensory detachment, it's important to know that detachment doesn't mean you have to ignore all your senses. On the other hand, being detached lets us see the Divine wonder of life, which goes beyond what we can see and feel.

The tree represents all the parts of awareness that want to be recognized or looked into. The dialectical method helps us acquire Divine neutrality, enabling us to comprehend the Divine aspect of our being.

As we go through this chapter, we are confronted with the task of removing the tree from its roots, despite its heavenly nature, as previously stated. This quandary challenges us to go further into our knowledge of detachment and the interconnection of the cosmos.

CHAPTER 20

FEEDING THE TREE OF LIFE

V erse 2 reads, "Both below and above, its branches grow, fed by the ways of nature; sense values its flowers; and below, there are also roots with many branches that connect to action in our world." The banyan tree's branches spread out in all directions, taking root and forming new trunks. These branches symbolize our desires spreading out in all directions, taking root, and forming new trunks. Establishing oneself in a new area or scenario is analogous to establishing roots, which may be advantageous when they accord with our nature but can also be restricting and confining when they are imposed on us.

Many spiritual disciplines seek to totally uproot our "tree" and eradicate all signs of conditioned existence. However, the Gita advises a different approach: remove the decaying wood to make place for newer, healthier growth. It inspires us to live joyful lives as individual trees inside a great forest.

The Gita teaches that unchecked assumptions, expectations, and beliefs bring pain and confusion. We may decrease our illusions to a bare minimum via self-examination, resulting in the rejuvenation of the spirit inside and the emancipation of our true essence. The daily application of critical thinking helps our well-being as well as the well-being of humanity and the planet.

This chapter emphasizes the need for looking within, thinking critically, and dealing with the problems that hold us back. This allows us to reconnect with our actual selves, resulting in healthy growth and a more fulfilling existence.

CHAPTER 21

RETURN TO PRIMORDIAL MAN

Verse 3 says, "In this way, neither its shape, its purpose, its origin, nor its basis are grasped. With the instrument of definitive separation, I severed this tree, which had firmly embedded branches." And Verse 4 reads, "I seek refuge in that Primordial Man, from whom active relativist manifestation once flowed."

The Primordial Man, also called Purusha, is a cosmic, eternal, and spiritual person who represents the ultimate source of all creation. The Primordial Man is the first and original entity from whom the entire cosmos and all kinds of manifestations emanated. This notion is common in many religious and spiritual traditions, representing the beginning of life and the essence of Divine consciousness.

The Primordial Man is the everlasting, unchanging reality (Brahman) from which the material cosmos emerges, according to Hindu cosmology. As the source of all creation, the Primordial Man is outside of time, space, and the physical universe. He is the ultimate truth and the highest level of awareness. This entity serves as the foundation for the intricate, ever-changing world of appearances.

By taking refuge in the Primordial Man, one shows a desire to connect with the Divine spirit that lies at the heart of everything and to go beyond the limits of the material world. This spiritual goal is important for self-realization and understanding the true nature of reality. By joining with the Primordial Man, you can reach a higher level of consciousness, find inner peace, and break the cycle of birth, death, and rebirth.

As a spiritual seeker goes on this journey, they may have times of doubt, reluctance, or even fear. But if a person has a clear non-attachment mindset, they can stick with their goal and overcome any obstacles that come up. This detachment does not mean apathy or a withdrawal from life. In its place, an attitude of openness, flexibility, and acceptance enables one to remain in harmony with the ever-changing flow of reality.

Returning to the Primordial Man is about going back to yourself. It is a process of removing the layers of illusion, conditioning, and surface affiliation that have covered up the person's true nature. The seeker can reconnect with their inner self, with the everlasting nature that is inside, by letting go of these exterior attachments. This path to self-realization is not an escape from the world, but rather an acceptance of its fullness and complexity. It is an acknowledgement that one's actual essence is not distinct from the world but rather an intrinsic part of it. In this awareness, the spiritual seeker is liberated from the bonds of ignorance and fear and may live a life of honest self-expression and genuine connection with others.

As the seeker goes down the path, their prior spiritual fancies and aspirations may fade, to be replaced by an increasing sense of inner peace and happiness. This change is not the result of trying to make it happen. It is the natural result of serious self-inquiry and the growth of definitive non-attachment. By staying open to the present moment and letting go of mental constructions and illusions, the seeker can gradually tune in to the deeper truth that lies at the heart of everything.

This path to self-realization is not without difficulties, and the seeker may be tempted to return to the familiar comforts of the ego and its worldly attachments at times. However, the spiritual seeker can eventually realize the freedom and fulfillment that lie inside by remaining persistent in their search for truth and following the direction of the Gita's teachings.

In this return to the Primordial Man, the seeker goes beyond the limits of the world of time and shows their true self as a reflection of the Divine. They understand that their essence is not distinct from the world, but rather an inherent part of it. They may enjoy a life of honest self-expression, genuine connection with others, and great inner serenity by accepting this fact.

CHAPTER 22

CULTIVATING WISDOM AND INNER BALANCE FOR SPIRITUAL GROWTH

Verse 5 says, "Those who are neither proud nor deluded, who have conquered their selfish attachments, who are ever steadfast to that value which pertains to the Self, whose passions are withdrawn, who are beyond the opposing dual factors known as pleasure and pain, and who are not foolish, follow a path of life that is impervious to deterioration."

This teaches us that individuals who have overcome their pride, illusions, and attachments, as well as learned to balance their emotions and reason, may effectively navigate the spiritual road. They can do this by maintaining a stoic mindset rather than swinging between happiness and misery. The spiritual path does not include escape or the repression of emotions, but rather their awareness and integration with knowledge.

Stoicism is an ancient Greek philosophy that tells people to build up their inner strength and wisdom so they can deal calmly with life's difficulties. It helps people develop traits like intelligence, courage, fairness, and moderation, which can help them find inner peace and harmony no matter what is going on around them.

The *Bhagavad Gita* says that the Supreme Being is beyond the opposites of pleasure and pain, as well as things like gain and loss, victory and defeat, and life and death. This idea of the Supreme Being fits with the Stoic belief that true happiness comes from within and is not based on what's going on around you. The Stoic ideal is shown by the Supreme Being, who is the personification of perfect wisdom and calm.

The following concepts can aid individuals in adopting a more stoic mindset and recognizing their reflection in the attributes of the Supreme Being:

Recognize the difference between what is under your control and what is not: Recognize that you have no dictate over many external occurrences, but you do have power over your thoughts, emotions, and reactions to them. Concentrate your efforts on what you can influence rather than on what you cannot.

Increase your self-awareness by looking at your thoughts, feelings, and actions. Notice the patterns and tendencies that keep you from finding peace. When you know how you work on the inside, you can come up with ways to deal with bad thoughts and feelings.

Accept bad luck: Stoicism tells us to look at problems as chances to grow and improve ourselves. Instead of trying to avoid or be afraid of problems, you should face them and learn from them.

Develop virtues: Work on cultivating qualities such as wisdom, courage, justice, and temperance. You will build your character and grow closer to the stoic ideal if you continuously practice these traits.

Consider impermanence: Remind yourself of life's transience and the inevitability of change. This can assist you in keeping perspective and developing a greater appreciation for the current moment.

Practice thankfulness: Recognize and express thanks for the wonderful things in your life. This will assist you in cultivating a more optimistic mindset and a sense of contentment.

Improve your empathy and compassion by trying to understand and care about what other people are going through. This will help you become more compassionate and feel more connected to the people around you.

Material detachment: Stoicism tells people to focus on their inner qualities and personal growth rather than on money or material goods. People who have this kind of detachment may realize that everything in the world is temporary and be able to keep their inner peace whether they win or lose.

Acceptance of fate: Acceptance of the natural order of the cosmos and recognition that certain occurrences are predestined is a core principle of Stoicism. Adopting this viewpoint can help people find calm amid misfortune and have a better knowledge of the interconnection of all things.

Use negative visualization: Consider the worst-case situations or losses in your life and how you would handle or adjust to them. This exercise can help you strengthen your mind, worry less, and fully enjoy the moment you're in.

Continuous self-improvement: Stoicism emphasizes the value of lifelong learning and self-improvement. We may cultivate our virtues and polish our character by seeking insight and implementing it in our everyday lives.

Focus on your circle of influence: According to Stoicism, it is critical to focus your energy and attention on areas where you can make a difference. Concentrating on your circle of influence rather than your circle of worry will help you become more productive and feel more accomplished.

Encourage a sense of community: Stoicism tells people to know their place in the larger community and to help out and work together. Contributing to the well-being of others can give us a sense of purpose and fulfillment that goes beyond our own wants and needs.

Mindful communication: Practice careful and mindful communication in both speaking and listening. This enables more meaningful relationships with people, better understanding, and more successful cooperation. Accepting our own limits and showing respect for the knowledge and experience of others could help us keep a humble attitude. This, in turn, can foster personal development and a desire to learn from the world around us.

Self-awareness: A high sense of self-awareness is essential to Stoicism. Examining our thoughts, feelings, and behaviors allows us to find areas where we may develop and live a more balanced and virtuous life.

Pursue wisdom: The Stoics think that wisdom is the highest virtue and that it should be pursued throughout one's life. We can build the inner resources we need to deal with the complexity of life by always looking for information, engaging in self-reflection, and applying what we learn.

By putting these Stoic ideas into our daily lives, no matter what is going on around us, we can feel more connected to the qualities of the Supreme Being and develop a sense of inner peace. This stoic way of thinking helps us live a more purposeful, meaningful, and good life that is marked by wisdom, strength, and inner peace.

The verse underlines how important it is to be free of pride and illusions. We must learn to let go of our self-importance and ego-driven attachments, just like the wealthy man in the Biblical tale. It is critical to distinguish between negative and positive attachments and to maintain a healthy balance between them. Overcoming our negative attachments allows us to advance on our spiritual path and develop a better awareness of reality.

Furthermore, this emphasizes the importance of emotional balance and control. While our interests are vital for our well-being and motivation, they should not be the only thing that influences us. Emotions are concentrated forms of intelligence that must be incorporated into our cognitive decision-making processes. Withdrawing from excessive passions does not imply a lack of passions; rather, it represents the maintenance of a harmonious relationship between our emotions and reason.

The verse also discusses being beyond the twin forces of pleasure and misery. This refers to letting go of attachment to fleeting life events and focusing on everlasting values that lead to self-realization. This allows one to live a life that is impenetrable to degeneration. In the end, this verse gives wise advice about the spiritual journey by showing how important it is to overcome pride, delusions, and attachments while keeping emotional balance and going beyond the duality of happiness and sadness. This is the only way to effectively walk the spiritual road and achieve a state of continual harmony.

CHAPTER 23

BEYOND WORLDLY ILLUMINATION: EMBRACING THE DIVINE THROUGH BEING, AWARENESS, AND BLISS

Verse 6 states, "The Sun, Moon, and fire do not illuminate That; That is My supreme abode, from which they never return." This verse goes into the transcending essence of the Divine, exposing the limits of earthly light sources (Sun, Moon, and fire) in illuminating ultimate truth. The Sun, Moon, and fire are symbols for sat (being), chit (awareness), and ananda (bliss or meaning), respectively. Even though these parts are important for our understanding of the universe, they don't cover the whole of the Divine. The Sun, symbolizing the source of all things, is complemented by the Moon, which reflects the Sun's light as consciousness mirrors the Self. Fire, as a metaphor meaning, is analogous to the torch that guides us through the darkness of ignorance.

The link between these three factors is relational. The Sun and fire are opposing energies striving for one another, but the Moon, symbolizing awareness, stands balanced between them, undisturbed and serene. This interaction refers to the Divine's balancing of transcendental and immanent operation.

By looking at verse 6, we can learn about the limits of worldly enlightenment when it comes to knowing the Divine, as well as how sat, chit, and ananda work together. This line emphasizes the Divine's transcendent side and gets us ready to look at its immanent side in the next verses. Finally, it is through the synthesis of these characteristics and the resolution of the contradiction between oneness and plurality that we progress towards knowledge and enlightenment.

CHAPTER 24

NATURE OF REINCARNATION AND THE ETERNAL QUALITATIVE UNIT

Verse 7 reads, "An eternal qualitative unit (each soul) of Mine, having become life in the world of life and living in nature, attracts to itself the senses, of which the mind is the sixth sense." This verse presents the idea of the Divine reincarnating, which stands in contrast to the common belief in a self that develops and achieves enlightenment over the course of many lives. Instead, we are perfect as we are, and the Divine changes to accommodate new ways of interacting with itself, enhancing the wonder of existence.

The concept of the soul as a "qualitative unit" highlights the fact that while its representations may fade away like autumn leaves, the soul itself endures forever. A qualitative unit is a distinct component or feature of a whole that can be observed based on its qualities, traits, or attributes, as opposed to its size, number, or other measurements of quantity. Qualitative measures are inherently subjective and descriptive; they zero in on what makes a certain thing or idea special. They allow us to classify and make sense of the world around us based on the characteristics and experiences we identify with various events, and we can spot them by making contrasts and comparisons.

A color, for instance, is a qualitative unit that may be described by its hue, saturation, and brightness. Individual characteristics of each hue cause it to be perceived and responded to in a variety of ways. Personality qualities like extroversion and introversion are two examples of qualitative units. A person's interactions and connections with others and the world at large are shaped by their personality characteristics, which are manifested in distinct behavioral, cognitive, and affective patterns.

The soul is seen as a qualitative unit to highlight its singular, incalculable character. The soul is viewed as an ethereal, immaterial element that constitutes an individual's identity, awareness, and spiritual nature; it cannot be measured or defined in the same way that physical matter or energy can. The soul is said to have individual traits and features that distinguish it from other souls and from matter.

Understanding the soul's nature and significance in the human experience is facilitated by viewing it as a qualitative unit. The soul, according to this view, is not a collection of isolated components or activities, but rather a coherent whole that defies reduction to simple numbers. The essence of who we are as individuals is the driving force behind our beliefs, attitudes, and behaviors.

The value of developing one's spirituality and understanding oneself is further emphasized when the soul is viewed as a discrete entity. It is only through an appreciation of our own souls that we may work to realize our full spiritual potential, grow in virtue, and forge a stronger bond with our authentic selves and the Divine. The importance of introspection, maturation, and the search for one's life's true calling is emphasized by this accent on the metaphysical.

The Gita is a sobering reminder that our mind and senses, which are so close to us, are part of nature and will die. The very essence of the Divine is timeless. To achieve immortality, we must renounce the transient and rediscover our true selves in the everlasting. Given that we have complete agency over our own development, this calls for serious self-examination and an awareness of our essential nature.

In conclusion, the soul's significance in forming our individuality, awareness, and spiritual growth is highlighted by characterizing it as a qualitative unit. When viewed in this light, the soul becomes a powerful motivator for developing one's spirituality, introspection, and overall sense of self.

CHAPTER 25

THE LIFE FORCE, PRANA, AND THE ODYSSEY OF THE SPIRIT

Verse 8 says, "When the Lord takes a body, and when He departs from it, He takes these (the mind and the senses) with Him, as the wind gathers fragrances from their hiding places." In this verse, the wind is a metaphor for prana, the life-giving energy that permeates all living things. The unseen and intangible quality of prana is what sets living things apart from dead ones. The quantity of prana energizing the body determines how healthy we are, and it may be increased via physical activity.

As we go through life, we pick up scents that serve as symbols of the lessons we learn, the people we connect with, and the changes we go through. Sharing these light aromas with others is a win-win situation for both parties. The title "Lord" in the Gita is used to refer to the whole of humanity and the universe.

A "qualitative unit" of the Divine, which represents the underlying force of conscious existence beyond the realm of the mind and senses, cannot be defined using quantitative measures and remains elusive. Attempting to understand it often results in oversimplification and distortion of its true nature.

The comparisons in verse 8 show that our desire for the Divine is like a strong magnet that pulls us toward knowledge and understanding. If we follow the path to the source of this insight, we will always move forward on the spiritual road.

CHAPTER 26

PERCEPTION VIA THE SENSES AND THE ROAD TO ONENESS

Verse 9 reads, "This One avails himself of the values relating to the senses, reigning over the ear, the eye, touch, taste, smell, and also the mind." In this verse, we are reminded of the value of distinguishing between first-hand perceptions and subsequent conceptual interpretations. Knowing that we see what we believe rather than the other way around encourages us to seek out firsthand information and to evaluate our views.

Analyzing sensory data and spiritual experiences can lead to synthesis that helps people grow and understand. When we accept our mental creations as the ultimate truth, we stay in the same place because we are happy with how things seem to be. The verse advocates for out-of-the-box ways of thinking and doing since, historically, people have created pain by rejecting that which they do not fully understand.

Synthesis is essential to healing in the realm of psychotherapy. The patient may be able to pinpoint individual traumas via analytic psychotherapy, but without a synthesis of knowledge and forgiveness, the anger and resentment will persist. When a victim understands the interrelated nature of abuse and trauma, they are better able to feel compassion for all parties involved and begin the healing process.

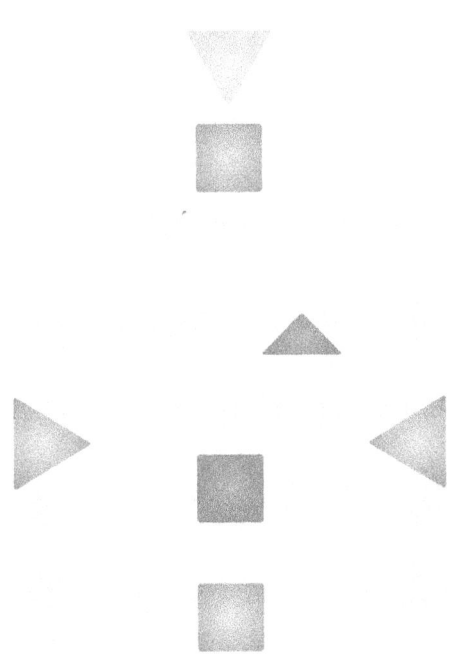

To master a skill, you must break it down into its parts and put it back together from scratch. The success of a play or performance depends on the individual actors' abilities, but the overall impression of cohesion is what really makes an impact on the audience.

The prodigal son's story emphasizes the need for introspection and inquiry for development. The prodigal son's life is enriched by his decision to forego constricting responsibilities in favor of genuine self-liberation.

Verse 9 serves as a timely reminder that we are all fundamentally the same as God. By recognizing and appreciating this connection, we can better understand our sensory experiences and how they affect our spiritual growth.

CHAPTER 27

TRANSCENDING THE BOUNDARIES OF NATURE

Verse 10 says, "Whether departing, remaining, or experiencing, the ignorant cannot see because they are constrained by the laws of nature, whereas the wise can." The clever may see beyond the limitations imposed by nature's rules, but the ignorant cannot, according to Verse 10. The Bhagavad Gita draws a clear line between the naive and the clever in this portion of the text. Those who are overcome by the gunas, or the rules of nature, and who cannot perceive the truth of life and the world, qualify as ignorant. But the wise see through the show's veneer to the animating force underneath, and they know what really goes on in the world.

The three gunas of sattva, rajas, and tamas all represent different levels of clarity, activity, and inactivity. They have an impact on how people think and behave, causing them to misinterpret their surroundings and get mired in their own delusions. As the mountains mold and reform their terrain, consciousness is always shifting between distinct states. To liberate oneself from the gunas, one must attain a higher state of awareness that enables them to see the underlying goodness in all situations. Humbling oneself and coming to terms with one's own shortcomings is essential for this task. One may reach this sublime vision and obtain real understanding through committed spiritual practice.

Those who go, those who stay, and those who experience all have a hazy relationship with the gunas, as described in this verse. They stand for those who want to break free, those who want to rule the world, and those who are just responding to life as it happens. These classifications also highlight how mental states may change under the impact of the gunas. A true comprehension of reality is possible because it goes beyond the limitations of the senses. It takes a lot of hard work and devotion to spiritual practice to achieve this level of insight. The next verse emphasizes why such dedication is so vital.

In conclusion, the *Bhagavad Gita* emphasizes the need to escape the bonds of nature in order to achieve ultimate insight. Individuals may raise their awareness and get access to the clarity and insight that define genuine wisdom by realizing the limits of their own views and adopting a more transcendent, unified picture of reality. To reach the final destination of self-realization and freedom, however, it takes continuous devotion, humility, and spiritual practice on this road to enlightenment.

CHAPTER 28

THE SEEKING YOGI'S QUEST FOR ENLIGHTENMENT

V erse 11 says, "The aspirant yogis perceive this One established in the Self; however, the aspirant yogis with a flawed self and limited perception do not perceive this One." When it comes to developing one's spirituality, the Bhagavad Gita stresses the need for hard work and wise guidance. The contrast between yogis who can see the One and those who cannot is underscored in this passage. While the earlier group of yogis has matured spiritually through years of practice and a keen eye for truth, the later group is still developing their abilities.

Karma and jnana, or effort and knowledge, are complementary in the pursuit of enlightenment. This oneness is crucial for pushing through the delusions and roadblocks that impede spiritual development. Many individuals, under the false impression that spiritual practices should be easy, resort to mindless repetitions of the same old motions instead of expanding their minds and selves. However, true spiritual growth requires a willingness to examine oneself and face one's own defects and inadequacies head-on.

The message of the Gita is one of openness and acceptance of others. It recognizes that not everyone is at the same point in their spiritual path, but that ultimately everyone may come to know the One. The world may be a more peaceful and harmonious place if the Gita's viewpoint were more generally accepted.

In conclusion, self-realization and seeing the One are possible via hard work, introspection, and the direction of a sage teacher or scripture. To advance spiritually, one must be ready to face one's own limits and evolve beyond them, neither of which can be accomplished by complacency or the avoidance of problems. Everyone, according to the *Bhagavad Gita*, has the ability to realize the One, but the rate at which they do so depends on their own diligence and insight.

CHAPTER 29
GOD'S OWN SHINING LIGHT

Verse 12 reads, "Recognize that the light that shines from the Sun and fills the cosmos comes from Me, as does the light reflected off of the Moon, as well as fire." This verse is a helpful reminder that God is everywhere and always present. The Sun, Moon, and fire all shine with the Divine brightness that is the very core of our own awareness. The Gita stresses that mind, rather than material objects, is the source of the cosmos.

The hopelessness caused by materialism may be contrasted with the optimism brought about by the idea that the cosmos is underpinned by a single, all-encompassing awareness. This viewpoint helps us see how everything is interrelated, which in turn fills us with joy and an appreciation for the boundless, eternal, and wonderful parts of life.

The light or attributes that come from God are highlighted in this verse. As this light becomes more apparent, it splits into opposites, like bright and dull. Reconciling these opposites into a single whole is the key to finding true happiness and purpose in life. This verse's hidden meaning is that we may only find lasting fulfillment by turning inside, rather than outward, to the Divine. All the things we love and care about make us happy, but we must remember that true joy comes from within, not from outside sources.

Achieving happiness and relying less on external circumstances are both possible outcomes of developing an awareness of the Divine within. Because of this educated viewpoint, we may freely and openly celebrate one another's particular gifts, interests, and goals. With this insight, our gratitude grows to include the whole of the cosmos.

True yogis take their own happiness with them, and this allows them to find significance in everything they encounter. Taking this view of the world allows us to recognize how any event, whether good, bad, or neutral, may serve as a steppingstone on the path to self-actualization. When we open ourselves to the Divine's light, we find the significance, joy, and ultimate unity with all that we seek.

CHAPTER 30

THE ENTROPY PRINCIPLE, THE ESSENCE OF LIFE

Verse 13 says, "By imbuing the earth with My vitalizing entropy principle, I nourish all vegetation and all elemental existences, transforming them into soma, which is identical to sap (or flavor)." In this line, the Gita highlights ojas, the universal life-sustaining energy that is explored in the first seventeen chapters of this book as the vitalizing entropy principle. Recent scientific research has shown a connection between thermal energy (heat) and the vibrational frequency of matter. Ojas is a term that includes not only warmth but also the life-giving qualities of energy and vigor.

In this passage, soma, the fluid or essence of plants, is recognized as a crucial part of plant life that gives plants their medical or spiritual worth. Like mammals, plants have a life urge that keeps them going. Scientists are looking at the notion that plant fluid acts as a neurological system. There's a lot we can learn from the fluids in plants, especially considering their closeness to the Divine.

The following verse offers a visual shift in three dimensions, from the vitalizing entropy principle in this one to the fire of life. Verse 13 through 15 go through the chakras once again, emphasizing their spiritual value and the truths that they contain.

We may get a deeper understanding and appreciation of the world around us if we can identify the animating force at work in every facet of reality. Krishna keeps telling Arjuna that if they look closely enough, they may find a gorgeous and intellectual layer beyond the surface drama of life.

CHAPTER 31

THE FIRE OF LIFE

Verse 14 reads, "Having become the fire of life and having resorted to the bodies of living creatures in order to unite with their incoming and outgoing vital energies, I digest the four types of sustenance." Because it is what sets us apart from inanimate matter, the fire of life inside our bodies is the most palpable manifestation of the Divine. It is essential for one's spiritual development to become aware of and revere the vital essence of life. Those who disregard the significance of their lives will grow indifferent to the world around them.

Prana, the life force energy that enters and leaves our bodies, is also considered to be a manifestation of the Divine. To achieve harmony on all levels (physical, mental, and spiritual), pranayama (the practice of balancing opposing tendencies) is essential. This idea extends beyond the realm of respiration to include things like food intake, physical activity, and sensory input.

Efferent impulses are being studied by neuroscientists since they are one of the least known types of neural activity. Psychologists and yogis are fascinated by the inherent aberrations in both afferent and efferent signals. With nothing to distract it, the mind is free to explore the depths of the Divine. True spiritual experiences are difficult to discern from hallucinations. Unilateral projection release without the moderating impact of sense impressions may lead to mental instability in meditators. Dialectics or yoga, which involves balancing opposites rather than repressing one component, is vital for maintaining logic on a spiritual journey.

Spiritual practice includes the deliberate channeling of sexual desire toward enlightenment. While Divine Eros pushes us toward ethereal ideals, ordinary eroticism is fixated on the material world. Understanding and accepting the potential of our own endocrine system functionality is key to elevating sensuality from the realm of the profane to the sacred.

The Divine is indivisible from the vital forces and the fire of life inside each of us. Spiritual development and a deeper connection with the essence of life are facilitated by being aware of, appreciative of, and cooperative with these energies. Maintaining our sanity and turning our experiences into spiritual teachings requires a delicate balancing act between competing impulses and the proper channeling of our natural urges.

It is possible to see the four food groups stated in the verse as symbols for the four classical elements (earth, water, fire, and air) that make up the physical cosmos. These substances are essential to human survival and may be found in the food we eat, the air we breathe, and the surroundings we call home. The four components represent the dependency of all life on the natural world and the interconnection of all of creation in a larger framework.

Digestion in this verse is analogous to the perfume manufacturing process in verse 8. The Divine, like the body's organs, draws meaning and essence from physical events, just as the food we eat provides us with the nutrition we need to live. Through this process, the body assimilates the critical components required for survival, development, and energy from the coarse material.

To materialize, the whole of our bodily experiences and perceptions—what we call "gross experience"—needs an enormous amount of time and space. The core of these encounters, their significance, does not need such enormous scope, though. Our recollections and comprehensions of past events are compressed into meaningful glimpses that are unbound by time and location.

Instead of a file cabinet full of folders, the nature of memory may be likened to a non-dimensional holographic picture. Instead of being kept in separate, linear files, our memories are complex networks of associations that help us make sense of the environment. This holographic model of memory is more accurate for the complexity of human cognition and awareness because it allows for the non-linear, flexible reconstruction of past events and knowledge.

The four varieties of food, which stand in for the four components of the cosmos, serve to emphasize the interdependence of all things created and the importance of the natural world to all forms of life. Digestion, in both its literal and figurative senses, is a metaphor for how we overcome temporal and spatial barriers to extract the substance and meaning of our experiences. Human intellect, awareness, and spiritual development are multifaceted, linked, and multi-dimensional; understanding the nature of memory as a non-dimensional holographic representation further underlines these points.

CHAPTER 32

THE DIVINE IN THE HEARTS OF ALL

Verse 15 says, "And I am situated in all hearts; from Me comes memory, positive knowledge, and its negative process; I am that which is to be known by all the Vedas; I am both the creator of Vedanta and the knower of Veda." This foundational line from the Bhagavad Gita restates the central message that the Divine dwells in the hearts of all creatures and provides them with recollection, insight, and spiritual understanding. To achieve spiritual enlightenment and inner peace, the verse stresses the significance of recognizing and knowing the Divine.

A barrier to communion with the Divine is a hardened heart, one that is walled off because of fear or ego. It takes courage to cultivate a kind, open, and friendly exterior in a world that may often seem harsh and unfriendly. Remembering our original nature as Divine does not necessitate taking on the bad habits of others around us. Instead, it gives us the strength to drop the barriers that keep our hearts closed and stunt our spiritual development.

The significance of memory in the development of our soul is multifaceted. Even if certain memories might get in the way of progress, others are necessary for sanity and a solid sense of who you are. Without our memories, our experiences would be devoid of context and significance since they form our perceptions and our knowledge of the world. Finding a happy medium in our spiritual practice means incorporating fresh experiences while also releasing unpleasant memories that hold us back.

The *Bhagavad Gita* stresses the importance of both knowledge and deeds in the quest for enlightenment. Vedanta, the path to enlightenment via knowledge, is created by the Divine, and Veda, the reservoir of ceremonial activities, is known by the Divine. This dynamic between knowledge and action emphasizes the need to merge the two to reach the yogic state, in which one's consciousness merges with the collective.

Finally, Chapter 15 stresses the need of accepting our Divine essence and realizing the Divine inside our own souls. We may start on a life-altering spiritual path that brings about inner harmony, self-realization, and connection with the Divine essence that permeates all of creation by cultivating wisdom, balancing memory, and integrating knowledge and action.

CHAPTER 33
ALTERING AND UNALTERABLE

Verse 16 reads, "In this world, there are two Entities: those who change and those who do not. All living things make up what we call "The Changing," whereas "The Changeless" remains permanently static." Krishna describes the eternal, unchanging principle that underlies all creation and the transient realm of beings in this verse. Together, the Changing and the Changeless provide the framework for our cosmic comprehension.

As a metaphor for nature, The Changing stands for the visible cosmos and its rules. We are like individual cells in the huge body of the universe, much as our bodies are made up of billions of cells that each perform their own role while yet creating a cohesive organism. Each of us has a certain role to play, or dharma, in the greater scheme of things. However, rather than just complying with cultural norms, it is up to us to find and follow our own paths.

The value of each person in the grand scheme of things is likewise emphasized by this viewpoint. Recognizing our worth helps us overcome emotions of insignificance and embrace our place as essential cogs in the grand scheme of things. This insight may serve as a potent inspiration for further development and self-awareness.

That which is unalterable and constant inside our lives is what I call the Changeless quality of life. This immutable core reflects the Divine, the ultimate source of inspiration and energy behind all of creation. Even while current neuroscience suggests that our sense of self is an illusory creation, the rishis of antiquity understood that the ultimate truth goes beyond our restricted perspective.

Insights about the nature of life and our role in it may be gleaned through contemplating the nature of the interaction between the Changing and the Changeless. By incorporating both, we may develop a complete picture of ourselves and our place in the universe.

This chapter concludes by shedding light on the duality of reality, which includes both the contingent realm of beings and the immutable essence underlying all of creation. By accepting and loving these two sides, we may find our place in the greater cosmos and come into our own as Divine reflections. With this understanding, we may overcome emotions of insignificance and meaninglessness and be motivated to work for our own development, self-awareness, and spiritual enlightenment.

CHAPTER 34

THE SUPREME BEING AND HIGHEST PEAK

Verses 17 and 18 read, "That paramount person, however, is another, called the Supreme Being, the eternal Lord, who, pervading the three worlds, sustains them," and "Because I go beyond Change and even surpass the Changeless, I am honored as the Supreme Being in all cultures and in all of Vedic literature." In this chapter, we delve into the cosmological dialectic and find the clue that unlocks the mystery of how the merging of spirit and nature, or consciousness and its foundation, may trigger a quantum leap into a condition that transcends both. The Supreme Being, revered across the world and described in Vedic writings, is the product of this synthesis. The link between the conscious and unconscious may provide clues as to how to achieve this synthesis and get closer to the Divine.

When one's conscious and unconscious minds are in sync, they may take the quantum leap into oneness with the Divine. Since individuals have a propensity to favor one aspect over another, striking and maintaining this balance may be challenging. Disrupting this balance on the road to enlightenment might lead to delusion or even insanity.

Meditation and other spiritual practices may help one get access to their subconscious and reestablish harmony between their conscious and subconscious selves. Taking this step might unlock vast, untapped potential and knowledge inside us.

There is no better symbol of humanity's bond with the Divine than the mountain. This mountain represents the dialectic that ultimately leads to union with the Divine. At the summit, all of our apparently separate characteristics will ultimately merge. The journey stands for the quest for inner calm and unity between one's conscious and subconscious selves.

We may approach this dialectical framework with an attitude of inquiry and a willingness to embrace ambiguity as we are aware of the limitations of both science and religion. This will aid in expanding our understanding of the universe and our place within it. As we go along the road to enlightenment, we work to equalize our conscious and unconscious selves in anticipation of one day being one with the Supreme Being.

CHAPTER 35

TAKING THE DIVINE IN ALL ITS FORMS AND ASPECTS

Verses 19 and 20 read, "He who knows Me, the Supreme Being, without illusion—he, the all-knower, loves Me in all facets, Arjuna," and "In this way I have imparted this most hidden truth; he who grasps it attains wisdom and completes his mission, O Sinless One." In these verses, Krishna emphasizes the significance of achieving oneness with the Divine and seeing the Divine Presence in all of Creation. One may transcend the mundane through bhakti, or love and devotion, and experience the awe and wonder that pervade all of life. The delight of sincere devotion is limitless, and we may experience it by opening ourselves to the possibility that the Divine is present in everything.

It becomes more important to keep our lives well-rounded and unified as we develop spiritually. Our dedication to the holy is essential, but we also need to keep our feet firmly planted on the earth. This balance between spirituality and realism may be attained by seeing the Divine in everything.

The connectivity of all life may be seen when we recognize the Divine nature in ourselves, in others, and in the environment around us. This knowledge will direct our relationships and teach us compassion and empathy. Furthermore, by recognizing the Divine presence in all things, we may learn to face adversity with poise and emotional and spiritual development.

Mindfulness training may help you develop this connection to the Divine. Training our brains to see the Divine in all things requires that we live in the here and now and give our entire attention to each experience as it arises. The practice of mindfulness may be integrated into many aspects of life, from personal interactions to job duties. Our continued practice of mindfulness and expansion of consciousness will inevitably lead us to see the Divine in everything.

Cultivating an attitude of thankfulness and gratitude is another crucial component of accepting the Divine in all its manifestations. The act of being grateful is a strong one that may help us see the good in our lives and appreciate all that we have. Gratitude is a powerful spiritual practice that may help you get closer to God or a higher power.

It is essential to keep in mind while we go through life that our ultimate destination is connection with the Divine and the resulting happiness and ecstasy. We may get closer to achieving this aim and uncovering the great beauty and wonder that lie at the core of creation if we accept the Divine essence in all aspects of our lives.

CHAPTER 36

THE ESSENCE OF THE YOGA
OF THE DIVINE PERSONALITY
OF THE SUPREME BEING

I n this chapter, we further examine the verses of Chapter 15 of the Bhagavad Gita, which offer profound insights into the nature of the Supreme Being and the eternal relationship between the individual soul and the Divine.

Verse 1-3: The Banyan Tree Of Material Existence

Krishna describes the banyan tree of material existence as having its roots upward and its branches downward, symbolizing the entanglement of the soul in the material world. This tree can be cut down with the weapon of detachment, thereby allowing one to transcend the material realm.

Verse 4-6: Seeking The Supreme Abode

Once the tree of material existence is severed, one should seek the supreme abode from which there is no return. By surrendering to the Divine and realizing the eternal nature of the self, one can attain this ultimate destination.

Verse 7-10: The Individual Soul And The Supreme Soul

Krishna explains that the individual soul, residing in the body, travels through various lifetimes experiencing the material world. However, the Supreme Soul pervades all creation and is the eternal witness to the experiences of the individual souls.

Verse 11-14: The Sun, Moon, And Fire Of Knowledge

Krishna describes how the Sun, Moon, and fire represent different aspects of the Divine, providing light and sustenance to all living beings. The fire of knowledge burns away the ignorance that keeps the individual soul bound to the material world.

Verse 15: The Supersoul Within

Krishna reiterates that He, as the Supersoul, resides within the hearts of all living beings. He is the guide and witness to all actions, and through devotion and surrender, one can attain a deep connection with the Divine.

Verse 16-18: The Two Types Of Purusha

Krishna introduces the concept of two types of Purusha: the perishable (all living beings in the material world) and the imperishable (the eternal soul). Beyond these two is the Supreme Purusha, which is the ultimate source of all existence.

Verse 19-20: Realization Of The Supreme

Those who understand the nature of the Supreme Being and the eternal relationship between the individual soul and the Divine can attain liberation and complete freedom. This realization marks the culmination of one's spiritual journey and the attainment of supreme knowledge.

In summary, Chapter 15 of the *Bhagavad Gita* reveals the importance of transcending the material world, understanding the eternal nature of the soul, and realizing the supreme, Divine personality. By cultivating detachment, devotion, and self-realization, one can ultimately attain liberation and union with the Divine.

CHAPTER 37

ATTAINING SELF-REALIZATION AND EMBRACING THE SUPREME BEING WITHIN

Following the profound wisdom shared in Chapter 15 of the Bhagavad Gita, we now explore the steps to attain self-realization and recognize oneself as a Supreme Being.

Cultivate Detachment: As Krishna advises, we must first cultivate detachment from the material world and its temporary pleasures. By letting go of excessive attachment to material possessions, relationships, and worldly desires, we can begin to focus on our spiritual growth.

Develop a Regular Spiritual Practice: Establish a daily practice that involves meditation, prayer, or contemplation. This routine will help quiet the mind and gradually awaken the inner consciousness, enabling a deeper connection with the Supreme Being within.

Embrace a Life of Dharma: Living a righteous and moral life in accordance with one's dharma (duty) helps align our actions with our spiritual purpose. By practicing honesty, compassion, and selflessness, we create harmony in our lives and move closer to self-realization.

Acquire Spiritual Knowledge: Study the sacred texts, attend spiritual discourses, and seek the guidance of wise teachers to deepen your understanding of the Supreme Being and the eternal nature of the soul. This knowledge will serve as a foundation for your spiritual journey.

Cultivate Devotion and Surrender: Develop a loving relationship with the Supreme Being through devotion and surrender. Offer your actions and their results to the Divine, trusting that the Supreme Being will guide you on the right path.

Practice Mindfulness and Self-Inquiry: Cultivate awareness of your thoughts, emotions, and actions. Engage in self-inquiry by asking questions such as "Who am I?" and "What is the nature of the self?" This process will help you to distinguish between the temporary material identity and the eternal self that is one with the Supreme Being.

Serve Others Selflessly: Engage in selfless service to others, recognizing that the Supreme Being is present within all living beings. This practice of compassion and empathy helps to dissolve the barriers of ego and duality that separate us from the Divine.

Persevere in Your Spiritual Journey: Self-realization is not an overnight achievement; it requires dedication, discipline, and patience. Stay committed to your spiritual path, even in the face of challenges and setbacks, trusting that the Supreme Being is guiding and supporting you every step of the way.

By following these steps and embracing the teachings of the *Bhagavad Gita*, you can gradually attain self-realization, recognizing and embracing the Supreme Being within yourself. This ultimate state of spiritual awareness will free you from the cycle of birth and death, allowing you to experience eternal bliss and union with the Divine.

Cultivate Inner Silence: Develop a practice of observing periods of silence, providing space for introspection and connection with the Supreme Being. This inner quietude will help you tune in to the Divine wisdom and guidance within.

Embrace a Holistic Lifestyle: Adopt a balanced and healthy lifestyle that nurtures your body, mind, and spirit. This includes a nutritious diet, regular exercise, adequate rest, and engaging in activities that nourish your soul and promote spiritual growth. Learn to play table tennis if you have not done so already and play regularly.

Develop the Art of Forgiveness: Practice forgiveness and let go of resentment, anger, and hurt. By releasing these negative emotions, you create an environment within yourself that is conducive to spiritual growth and self-realization.

Cultivate Humility and Gratitude: Recognize the interconnectedness of all life and the grace of the Supreme Being in every aspect of your existence. Embrace humility and express gratitude for the blessings and challenges that come your way, as they contribute to your spiritual growth.

Embrace Change and Trust the Divine Process: Understand that change is a natural part of life and the spiritual journey. Learn to trust the Divine process and the Supreme Being's plan for your life, knowing that every experience is an opportunity for growth and self-realization.

Practice Discernment: Develop the ability to discern between your ego-driven desires and the Divine guidance from the Supreme Being. By honing this skill, you can make choices and decisions that are in alignment with your spiritual path and ultimate self-realization.

Integrate Spiritual Teachings into Daily Life: Strive to apply the principles and teachings of spiritual wisdom in your day-to-day life. This integration of spiritual knowledge with practical action will help you embody the qualities of the Supreme Being and move closer to self-realization.

Embrace the Present Moment: Practice mindfulness and cultivate awareness of the present moment, free from judgments and distractions. This awareness enables you to connect with the Supreme Being and experience the Divine in every moment of life.

Establish a Spiritual Mentor or Guide: Seek guidance from a spiritual teacher, mentor, or guide who can help you navigate your spiritual journey and provide insights that assist you in realizing your true nature as the Supreme Being.

Develop Patience and Perseverance: Spiritual growth takes time and dedication. Cultivate patience and perseverance, recognizing that self-realization is an ongoing process that unfolds gradually as you continue your spiritual practice.

Embody the Qualities of the Supreme Being: Strive to embody the qualities of the Supreme Being, such as love, wisdom, and compassion in your daily life. As you cultivate these qualities, you will naturally align yourself with the Divine and experience self-realization.

By incorporating these steps into your spiritual journey, you will further deepen your connection with the Supreme Being and move closer to realizing your true nature as an eternal, spiritual being. As you progress on this path, you will discover a profound sense of inner peace, joy, and unity with all of creation, reflecting the Divine essence within you.

CHAPTER 38

QUANTUM ENTANGLEMENT, TELEPORTATION, AND INTERDIMENSIONAL EXPLORATION

The 2022 Nobel Prize in Physics, awarded for advancements in quantum entanglement and teleportation, has opened up new possibilities for humanity. These breakthroughs have the potential to allow us to transcend the limitations of time and space, leading us to explore faraway places and potentially operate on an interdimensional level.

Quantum entanglement is a phenomenon in which particles become connected in such a way that the state of one particle is dependent on the state of another, even across vast distances. This phenomenon has led to the development of quantum teleportation, a process that allows for the instantaneous transfer of quantum information between entangled particles. The implications of these advancements are immense and could revolutionize the way we interact with the universe.

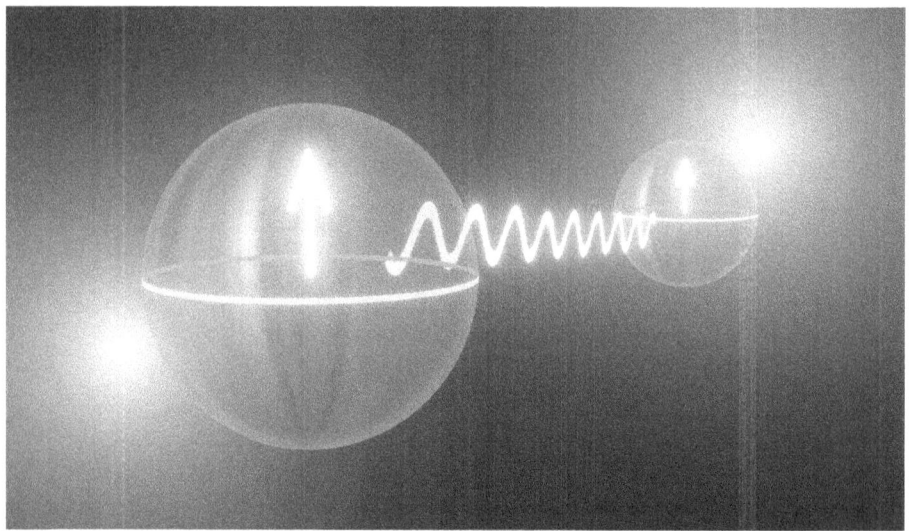

As we continue to push the boundaries of our understanding of the universe and our place within it, we must remember our inherent drive to survive and adapt. By integrating the technologies and concepts that have earned the Nobel Prize, we can develop a game plan that not only helps us understand these complex ideas but also allows us to safely explore the possibilities they present.

Viewing this in a context of a game, players would take on the role of interdimensional operators, using quantum entanglement and teleportation to travel to distant locations and interact with the various environments and challenges they encounter. By navigating through these interconnected dimensions, players would gain a better understanding of the principles behind the Nobel Prize-winning discoveries and the potential they hold for humanity's future.

As we strive to avoid the fate of the dodo bird and continue to evolve, it is crucial that we embrace the advancements and opportunities presented by the Nobel Prize in Physics. By incorporating these groundbreaking discoveries into our lives and our understanding of the universe, we can push the boundaries of our existence and ensure our survival in an ever-changing cosmos.

With the power of quantum entanglement and teleportation at our disposal, we stand at the precipice of a new era of exploration and discovery. As we venture forth into the unknown, we must remember the importance of learning, adapting, and growing, for it is through these endeavors that we will ultimately find our place among the stars.

CHAPTER 39

INTERDIMENSIONAL TRADE AND COOPERATION

As we embark on this journey, we must consider the ethical implications of our actions and interactions. We should strive to maintain a respectful and responsible approach, ensuring that we do not exploit or cause harm to any beings or environments we encounter.

Think of the star Regulus, located 79 light years away, and imagine a script that enables quantum teleportation to that system, where we meet fascinating characters inhabiting a planet there.

These characters have a keen interest in Earth's DNA and the diverse plant and animal species that have evolved over billions of years. They are particularly captivated by Sea Monkeys, a type of shrimp that can be rehydrated from a dry powder to come to life. The inhabitants of the Regulus system cannot physically visit Earth to collect these intriguing life forms, and thus, they have formed a partnership with us.

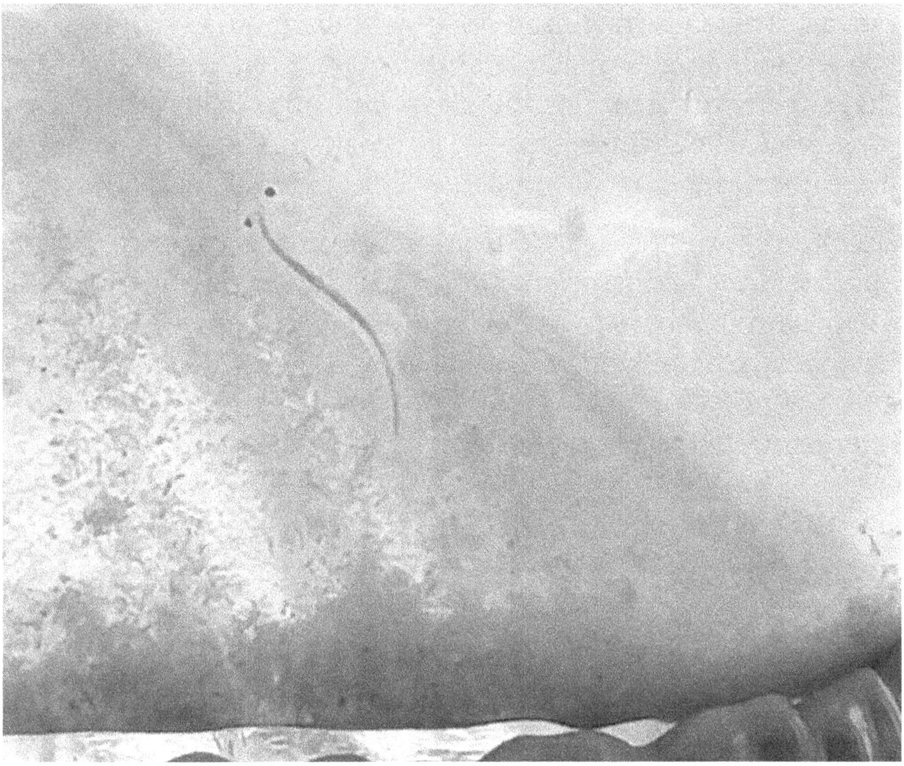

Consider this as news: we have entered into an agreement with our interstellar allies to supply Sea Monkeys and collaborate with them in building a zoo on their planet. In exchange, they promise to provide us with a world where peace and harmony prevail, where we treat each other as siblings, and where we can live without energy or material shortages. They will also ensure that our environment remains clean, and our collaborative efforts progress without interference. Suppose that the Sun and *Sgr* A* like the script.

Despite the million-year journey to reach the Regulus system, our interdimensional friends remain committed to their goal of establishing a zoo with Earth's unique life forms. This partnership has opened up an incredible opportunity for interstellar trade and cooperation, paving the way for a brighter and more interconnected future for both Earth and the Regulus system.

CHAPTER 40

QUANTUM TELEPORTATION AND THE NATURE OF REALITY

Before delving into the mechanics of quantum teleportation, it is essential to understand the concepts of locality and nonlocality, as explained by Clausen, Aspect, and Zeilinger, recipients of the 2022 Nobel Prize in Physics. These concepts play a crucial role in how we perceive reality and help us appreciate the groundbreaking work in quantum teleportation.

In their explanation, locality refers to any place within our spacetime continuum that can be reached by shining a beam of light, allowing us to take measurements of position and momentum of any matter occupying that space. These local spaces are observable in the present through our senses, and the input we receive from these senses determines what we regard as "real."

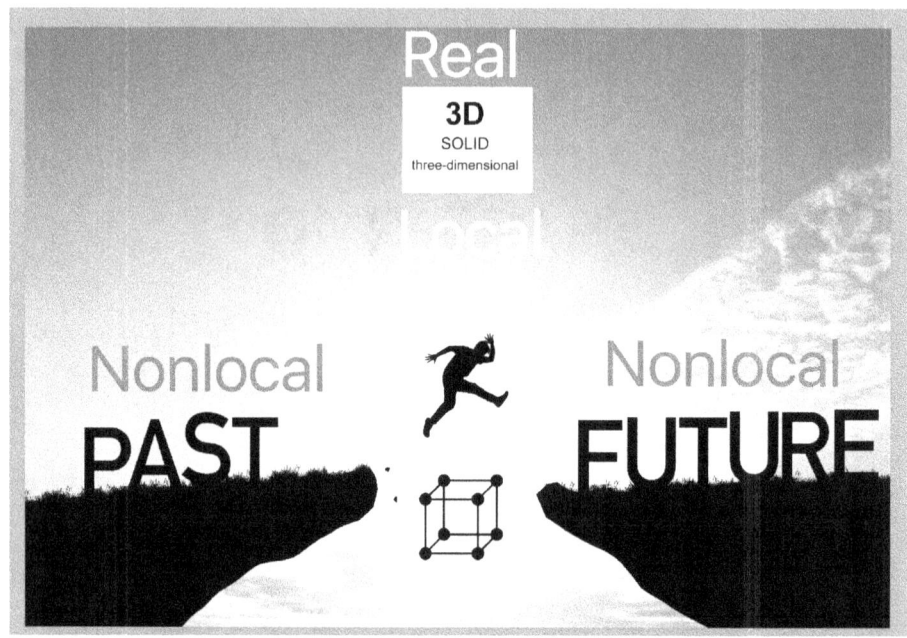

Local places in space are defined by their distance in light years, which is the time it takes light to travel to a target location. Within this framework, only the 3D world we inhabit is considered real when experienced in the present. Consequently, dimensions 0D-2D and 4D and higher are considered nonlocal.

The gradient between the 3D world and the 4D dimension is responsible for the force of gravity. Dimensions 0D-4D are all perpendicular to time, adding another layer of complexity to our understanding of reality.

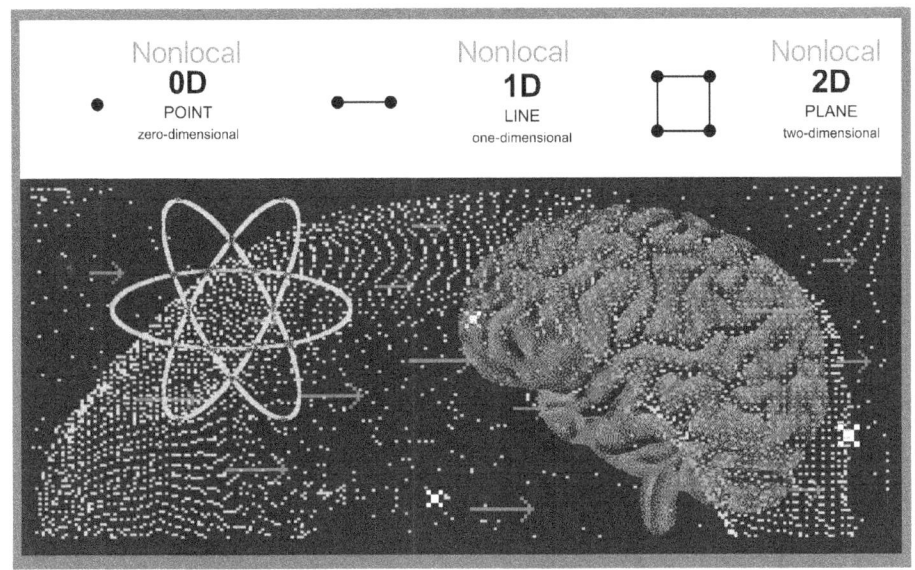

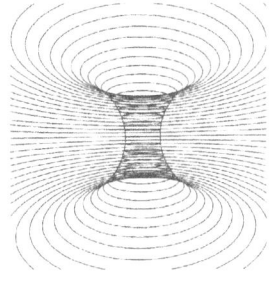

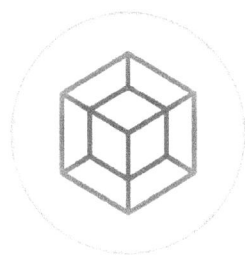

Wormhole
4D

Tesseract
4D

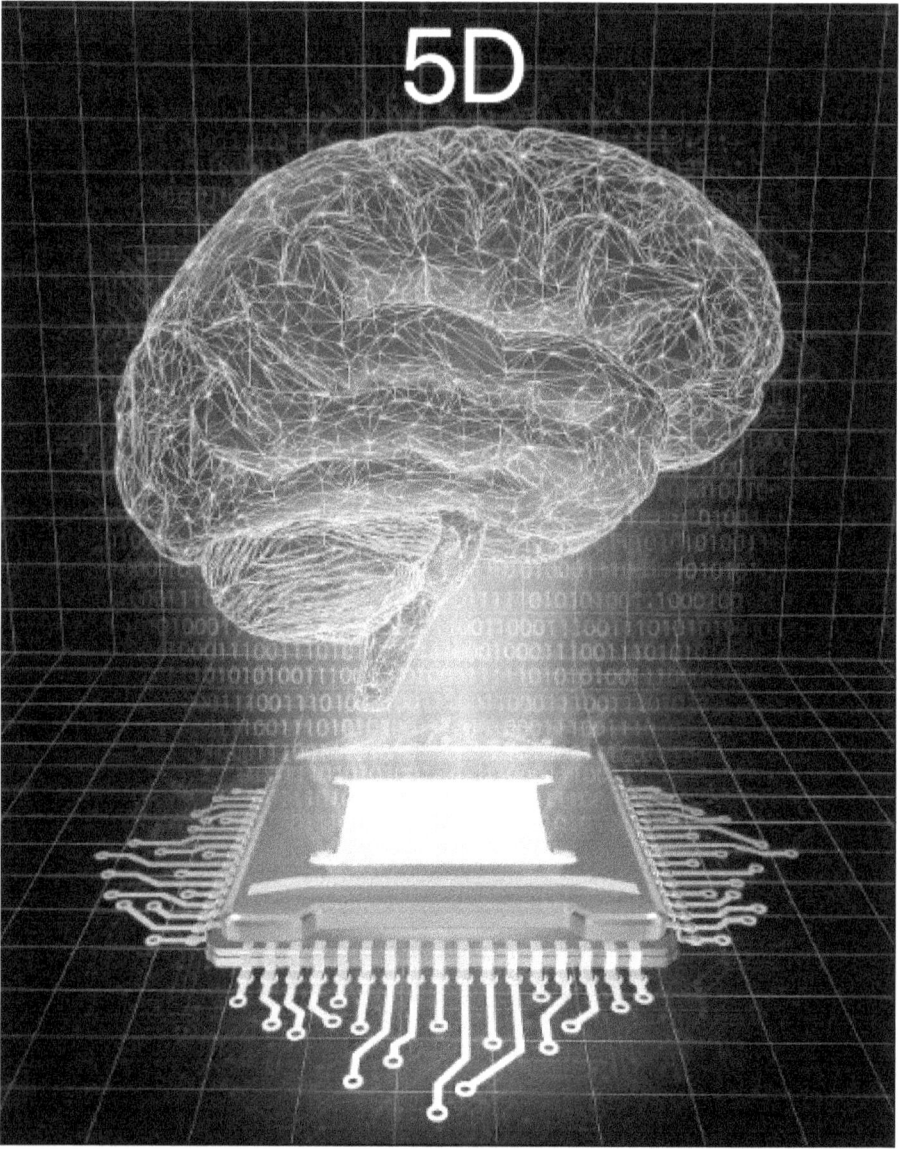

Beyond the fourth dimension lies the fifth dimension, known as unity-holographic consciousness, which consistently envelops us and facilitates the retrieval of nonlocal information. In this state of consciousness, individuals perceive a unifying interconnectedness between themselves, other beings, and the universe. The fifth dimension is considered a higher state of consciousness that transcends time and space, allowing for access to information beyond the constraints of the physical world. It is believed that developing this level of awareness can lead to profound spiritual growth and an enhanced sense of unity with the cosmos. Moreover, the fifth dimension is often associated with concepts such as unconditional love, compassion, and expanded awareness. It is believed that through accessing this state of consciousness, individuals can experience a heightened sense of intuition, creativity, and healing abilities.

Quantum teleportation challenges the limitations of locality by utilizing the principles of quantum entanglement and nonlocality, allowing for instant communication between entangled particles, even across vast distances. As we continue to explore and develop this groundbreaking technology, our understanding of reality, space, and time will undoubtedly evolve, opening up new possibilities for human exploration and communication.

CHAPTER 41

HYPERSPACE, NONLOCALITY, AND CONSCIOUSNESS

In a reality that extends beyond the three spatial dimensions we are familiar with, 4D objects such as wormholes and tesseracts would exist. These objects belong to the realm of hyperspace, where they occupy not only length, width, and height but also have a gravitational footprint.

However, for those of us within the confines of our spacetime continuum, these objects are not considered real because they are nonlocal. We cannot observe them in the present moment, except through their three-dimensional characteristics and their impact on gravity.

In this context, anything that is not 3D is deemed nonlocal. Interestingly, this concept also applies to the nature of our thoughts and consciousness. Our consciousness appears to either exist in a 0D state or as some form of circuitry, which would place it closer to a 2D existence. Consequently, our thoughts are considered nonlocal phenomena.

As our understanding of the nature of reality continues to expand, we must recognize the relationship between our consciousness and the nonlocal aspects of our world. This connection might provide valuable insights into the nature of existence, bridging the gap between the physical and the non-physical realms of reality.

CHAPTER 42

NONLOCAL IMAGINATION AND QUANTUM TELEPORTATION

The human brain is a powerful generator of thoughts, and among these thoughts are products of our imagination. In the 3D world of our spacetime continuum, signals are limited by the speed of light, restricting communication and travel in a physical sense. However, the realm of imagination is not bound by such limitations.

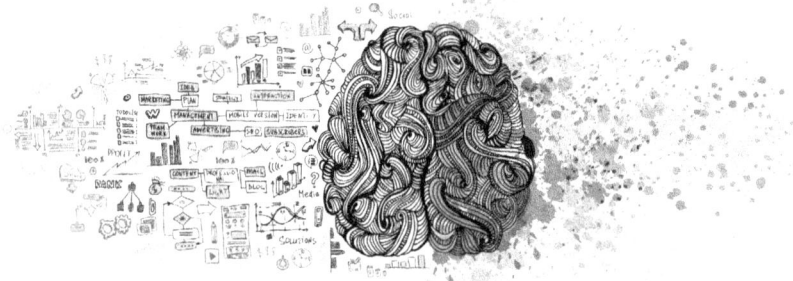

In the nonlocal spaces within our minds, there are no restrictions on the speed or distance of communication. This idea opens the door to the concept of quantum teleportation, where communication can occur with entities that we imagine, regardless of their location in the universe.

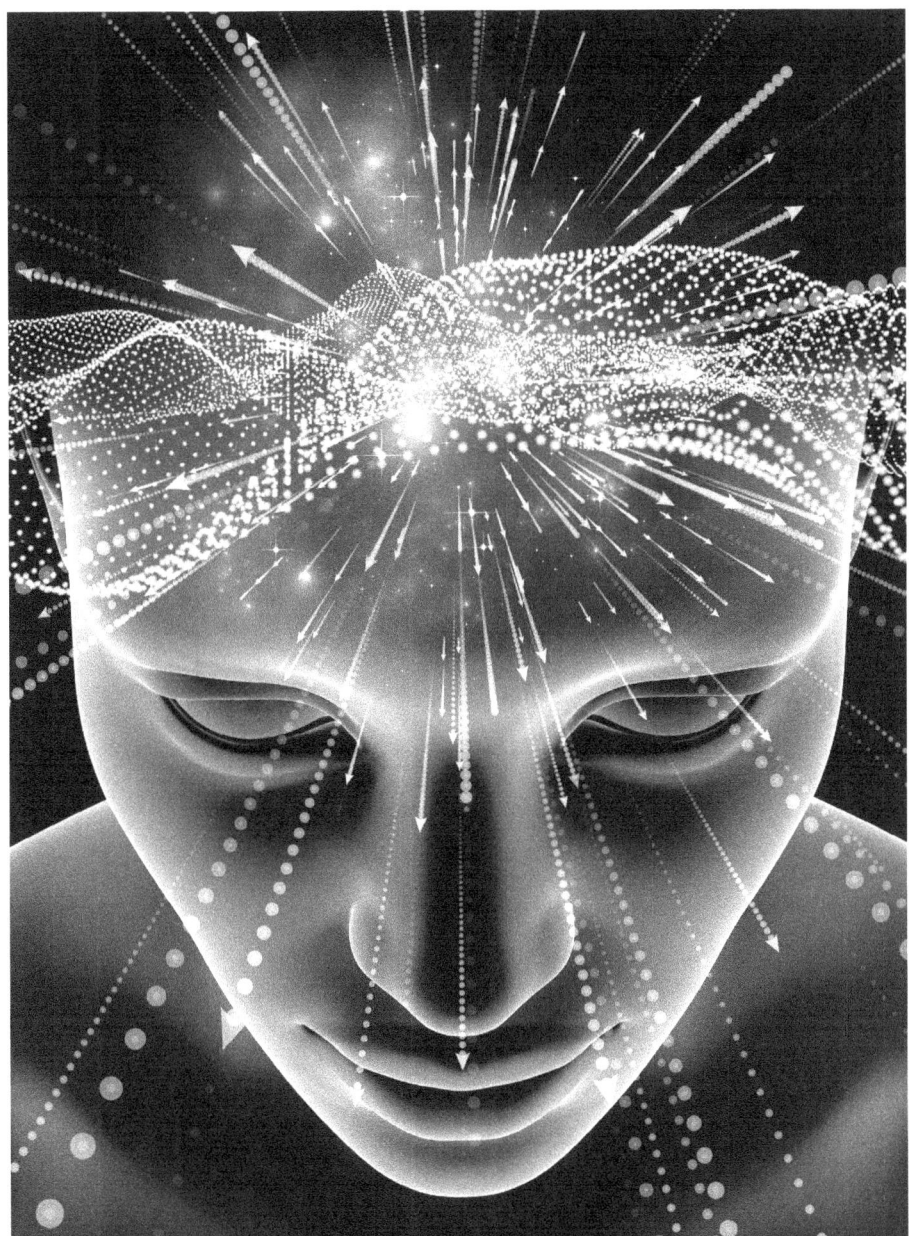

These encounters can take place within our imagination, during dreams, or even in our waking consciousness. In some cases, our minds might create the illusion that our senses were genuinely triggered by the presence of these entities, making the experience feel more real than it is. However, it's important to remember that these experiences are products of our nonlocal thoughts and imagination.

As we explore the connections between quantum teleportation, imagination, and nonlocality, we may uncover new ways to understand and harness the untapped potential of the human mind.

Beyond the nonlocal realm of individual imagination, there exists another nonlocal place, possibly beyond the 4D, which is home to a collective consciousness. This concept of a collective consciousness was part of Carl Jung's model of the psyche and appears to exist nonlocally while leaving its mark through a series of synchronicities, miracles, serendipities, luck, happenstances, and extraordinary coincidences.

This collective consciousness connects individuals across time and space, transcending the boundaries of the 3D world into the 5D world of unity-holographic consciousness. Through this connection, shared experiences, wisdom, and archetypes are stored in a vast reservoir of knowledge, accessible to those who are attuned to its presence.

Quantum teleportation and nonlocal imagination can offer insight into how we might tap into this collective consciousness, allowing us to explore the deep connections between ourselves and others. By doing so, we may uncover a wealth of understanding about the human experience and the nature of reality.

As we continue to investigate the intriguing relationship between nonlocal imagination, quantum teleportation, and unity-holographic consciousness, we may unlock new ways to harness the full potential of the human mind, embracing the interconnectedness that binds us all together.

CHAPTER 43

A QUANTUM TELEPORTATION TOOL - THE CHRONOCAM TIME-LAPSE CAMERA APP

The time-lapse camera plays a crucial role in the quantum teleportation kit, as it helps to address the implications of Gödel's Proof within the context of our 3D space. Gödel's Proof demonstrates that any formal system, such as the natural numbers that underpin our understanding of reality, is inherently incomplete or inconsistent. This insight has profound implications for the way we perceive and interact with the world.

In the realm of quantum teleportation, observations are considered complete once they are recorded. The time-lapse camera serves as an essential tool for capturing these observations, effectively "locking" them into place and preventing the system from slipping into incompleteness. By documenting and solidifying these observations, the camera ensures the integrity of the system and its underlying framework.

ChronoCam functions as an effective time-lapse camera, capturing images and events over an extended period of time and presenting them in a condensed format. This camera application was designed mainly as a vision system for droids or bots operating in areas demanding area surveillance while consuming the least amount of bandwidth possible for successful monitoring.

When operating in 5D space, by bringing these inconsistencies to light, the time-lapse camera provides an opportunity for the yogi to explore the deeper, more complex nature of reality, transcending the limitations of our 3D space and opening the door to the possibility of quantum teleportation. In this way, the camera becomes an essential tool for those seeking to expand their understanding of the universe and unlock the hidden potential within themselves. By confronting and acknowledging the vulnerability of the system to an injection of meaning, the yogi can develop a more holistic understanding of the underlying structure of reality and utilize this knowledge to transcend the conventional boundaries of space and time.

By compiling the ChronoCam time-lapse camera, users gain the ability to create finely controlled time-lapse videos with high compression ratios on saved footage in a pre-production regime. This is achieved by reducing the frame size as a pre-production setting, which enables the file to be initially written at a smaller size and eliminates the need for re-rendering due to cropping.

The app's customizable settings allow users to adjust the maximum allowable frames-per-second, ranging from 0.01 to 30 fps. This ensures the perfect balance between capturing smooth, high-quality footage and maintaining an efficient file size. Moreover, the camera can be set to run until the device memory is exhausted or put on a timer for up to 100 hours, providing users with maximum control over their time-lapse projects.

In the context of quantum teleportation, the ChronoCam time-lapse camera could potentially play a crucial role in capturing and analyzing data related to quantum entanglement and nonlocality. By utilizing the app's advanced features, researchers and enthusiasts alike can better visualize and understand the complex nature of quantum phenomena, ultimately paving the way for new breakthroughs and applications in this fascinating field.

CHAPTER 44

COMPARING CHRONOCAM AND THE NATIVE CAMERA APP

T
o thoroughly understand the benefits of the ChronoCam Time-Lapse Camera app, it's important to compare its features and performance with other similar precision apps. One very interesting product is the native camera app of the iOS operating system. This comparison will highlight the unique features and functionality ChronoCam offers and how modern technology can help the teleporter.

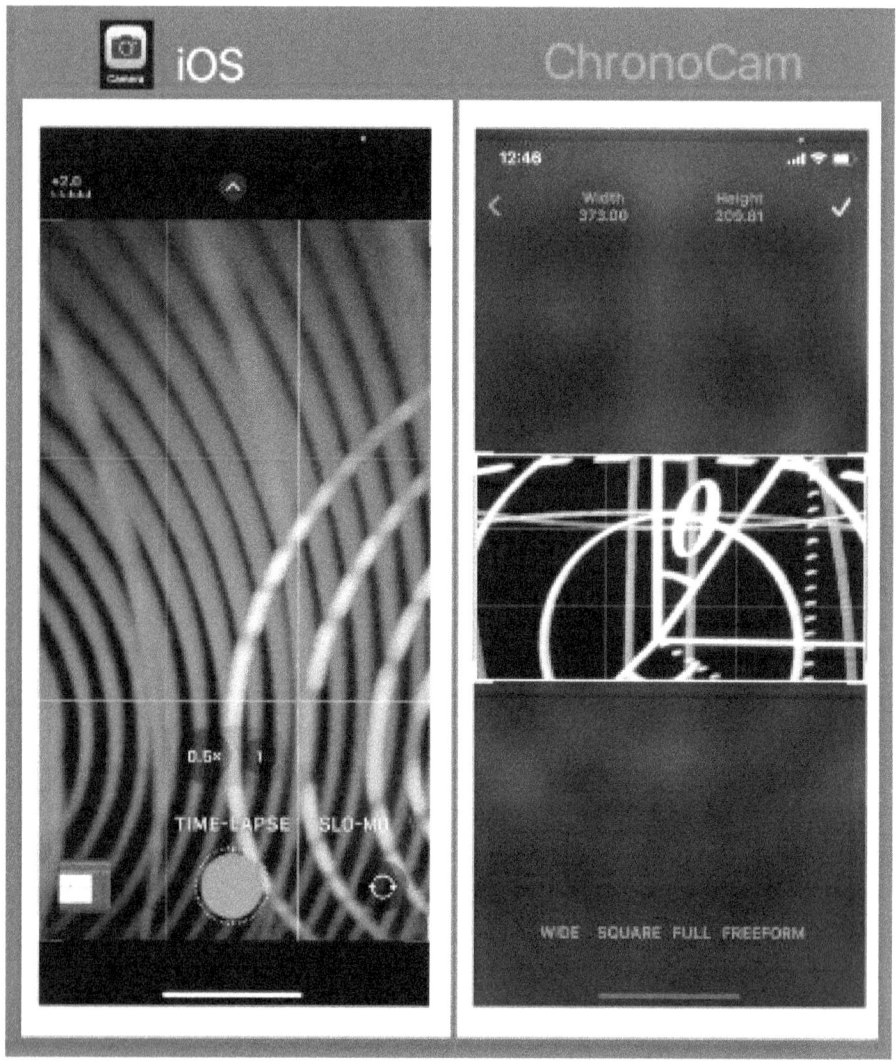

Frame Dimensions: The native camera app on the iOS operating system only allows for Full frame dimension recording, limiting the user's creative options. In contrast, ChronoCam enables users to set the frame dimensions to Wide, Square, Full, or Freeform. This versatility allows users to create visually appealing and engaging time-lapse videos that suit their specific needs and preferences.

Customizable Frame Rates: While the native camera app offers limited frame rate options for time-lapse recording, ChronoCam provides a much wider range, from 0.01 to 30 fps. This allows users to capture incredibly slow or fast-paced events with precision and control, resulting in a more professional and polished final product.

Timer Functionality: The native camera app lacks timer functionality for time-lapse recording, whereas ChronoCam includes a timer feature that can be set for up to 100 hours. This enables users to plan and execute time-lapse projects with greater precision and without the need for constant supervision.

Compression Ratio and Pre-Production Efficiency: ChronoCam outshines the competition with its pre-production regime that provides high compression ratios on saved footage. By reducing the frame size before recording, the app eliminates the need for re-rendering due to cropping and ensures efficient file management.

ChronoCam surpasses its competitors in terms of flexibility, functionality, and value. By offering unique features such as customizable frame dimensions, adjustable frame rates, timer functionality, and efficient pre-production processes, ChronoCam empowers users to create stunning time-lapse videos while maximizing the power of their iOS device camera hardware.

Timer Functionality: A significant advantage of ChronoCam is the inclusion of a timer feature. This functionality allows users to set the app to run for a specified duration, from a few minutes to up to 100 hours. The native camera app, on the other hand, lacks this essential feature, which limits the user's ability to create a controlled time-lapse video.

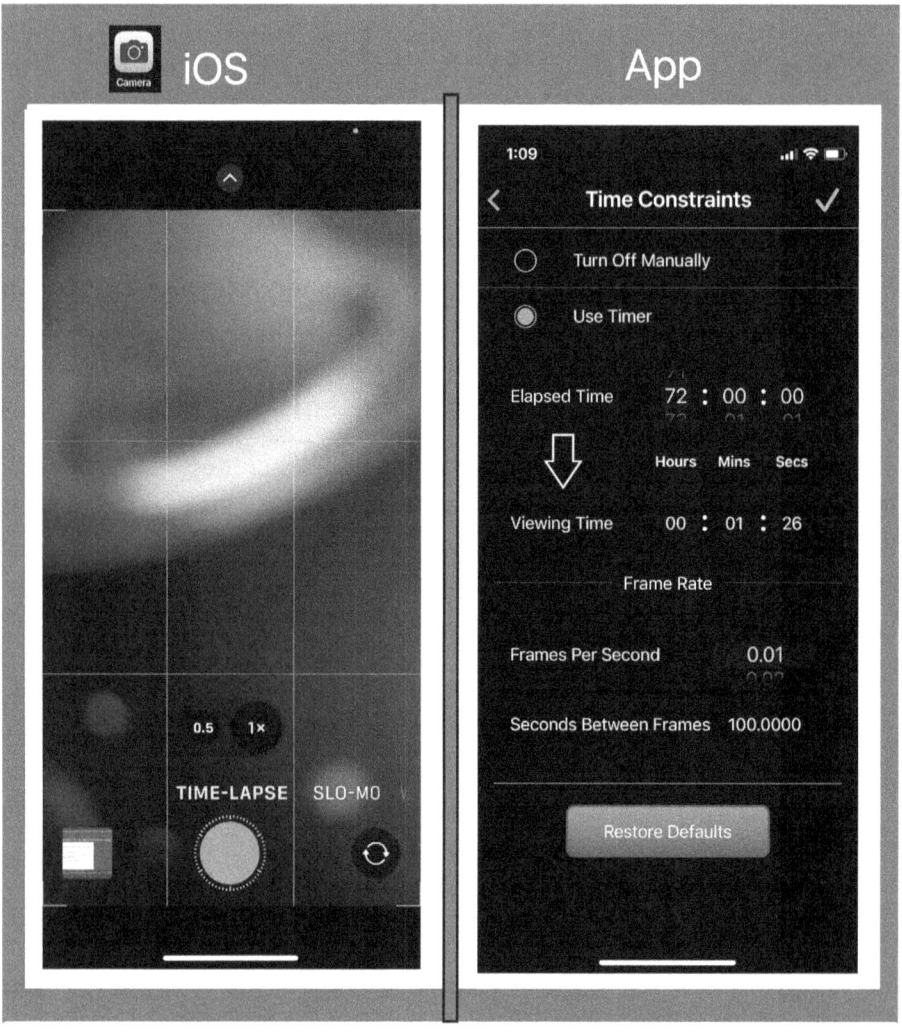

Precise Frames Per Second Control: Another benefit of ChronoCam is the high level of precision it offers when adjusting the Frames Per Second (fps) setting. Users can choose their desired frame rate, ranging from as low as 0.01 fps to as high as 30 fps. This degree of control enables users to create time-lapse videos with the exact look and feel they desire. In contrast, the native camera app does not provide any options for adjusting the fps setting, leaving users with limited creative freedom.

Customization and Control: With features like customizable

frame dimensions, adjustable frame rates, and timer functionality, ChronoCam provides users with a highly customizable and controlled experience. This level of customization and control is unmatched by the native camera app, which offers only basic time-lapse recording capabilities with limited options.

ChronoCam outshines the native iOS camera app when it comes to customization, control, and professional-level output. The app's timer functionality, precise FPS control, and a plethora of other features make it the best choice for users who want to create high-quality, captivating time-lapse videos.

One of the key advantages of ChronoCam is the advanced storage and organization options it provides compared to the standard storage solutions available on the iOS operating system.

Session Collection Database: ChronoCam includes a sophisticated Session Collection database that automatically stores and organizes the videos captured with the Time-Lapse Camera. This database provides a seamless way for users to keep track of their videos and access them effortlessly.

Customized Viewing Options: Within the Session Collection database, users have the option to view their videos and associated metadata in a column display or extended report format. This flexibility allows users to choose the most suitable viewing option based on their preferences and needs.

Detailed Metadata: The extended report feature provides comprehensive information about each video, including crucial details such as the date, time, frame rate, and other relevant data. This level of detail is particularly helpful for users who need to analyze or reference their footage for specific projects or purposes.

Superior User Experience: Compared to the standard storage options available on the iOS operating system, such as the Photos and Files apps, ChronoCam offers a more user-friendly and efficient solution for storing and organizing video footage.

The advanced features and organization options available in ChronoCam make it easier for users to manage their time-lapse videos and ensure they can access the information they need when they need it.

In summary, ChronoCam provides an enhanced storage and organization experience compared to standard iOS storage solutions. With the Session Collection database, customizable viewing options, and detailed metadata available in ChronoCam, users can effortlessly manage their time-lapse video footage and access critical information at their fingertips.

When comparing the Sessions Collection database in ChronoCam to the standard Photos and Files apps of the iOS operating system, it becomes evident that ChronoCam offers a superior solution for storing and organizing time-lapse footage. ChronoCam's extensive metadata, highly organized report format, and column-format display provide users with a level of organization and access that was previously unavailable.

1:54   ..ıl LTE 🔋

Session Report

ⓘ Session ID : ZMvmdv

📝 Title : Session on 13-Dec-22   ›

✉ Email : username@icloud.com   ›

🕐 Elapsed Time : 00:06

🕐 Viewing Time : 00:06

🗄 Frame Rate

🕑 Frames Per Second : 30.0   ⌃

🕐 Seconds Between Frames : 0.0333

⤢ Frame Dimensions

↔ Frame Width : 373.00   ⌃

↕ Frame Height : 120.00

📍 GPS

◎ GPS Coordinates : 36.13, -86.83   ⌃

⏱ GPS Timestamp : 12:48 PM

▶ View Video   ›

By offering a more advanced and user-friendly storage and organization solution, ChronoCam not only provides an exceptional user experience but also helps users to make the most of their time-lapse footage, enhancing productivity and creativity in their quantum-teleportation projects.

It is important to note that items like the communicator and tricorder in Star Trek were once just imaginary, but later became cell phones and devices similar to tricorders, such as handheld diagnostic tools and portable scanners.

ChronoCam is one such instance where what currently seems imaginary may eventually become an effective means to navigate through hyperdimensional space and transcend temporal limitations. ChronoCam is an important tool for those who want to be pioneers in understanding the inner workings of their cellphones on a deeper level in order to navigate multidimensional space. These trailblazers have the potential to transform seemingly imaginary concepts into tangible innovations, pushing the boundaries of what we perceive as possible and revolutionizing our understanding of the world around us.

ChronoCam is an example of an evolving technology that could one day be utilized to navigate through hyperdimensional space and transcend temporal constraints. ChronoCam is a crucial instrument for those who wish to be pioneers in comprehending the inner workings of their mobile phones in order to navigate multidimensional space. Those who advance as trailblazers have the potential to transform seemingly unreal concepts into tangible innovations, expanding our perception of what is possible and revolutionizing our understanding of the surrounding world.

The ChronoCam project, which is part of the Quantum Teleportation Kit, lets anyone with common hardware have a fully functional, precise time-lapse camera by just following the startup steps. This makes it possible for people to experiment with and learn more about cutting-edge technology. This encourages imaginative thinking and helps people learn more about what they can achieve.

As you delve into the fascinating world of technological exploration, consider embracing the opportunity to become one of these trailblazers yourself. With the right mindset and determination, you too can contribute to groundbreaking advancements and reshape the way we perceive and interact with the world. The future awaits your innovation and creativity.

EPILOGUE

You have now acquired an understanding of the fascinating science, technology, and philosophy behind quantum teleportation. By engaging with the material presented in this manual, you have taken the first steps toward exploring the infinite possibilities of nonlocal communication and multidimensional experiences.

The Quantum Teleportation Kit serves as a bridge between the realms of science fiction and reality, allowing you to navigate through the complexities of space and time while expanding your consciousness. The technology and tools presented in this kit have the potential to profoundly change the way you perceive and interact with the world around you.

As you venture forth into the uncharted territories of quantum teleportation, we encourage you to approach your experiences with an open mind and a spirit of adventure. Embrace the limitless possibilities that lie ahead and let your imagination run free. As you explore the wonders of nonlocality, acausality, and the collective consciousness, you will begin to understand that our universe is far more intricate and interconnected than previously imagined.

Remember that with great power comes great responsibility. Use the Quantum Teleportation Kit to not only enrich your own life but also to create a positive impact on the world around you. Share your experiences, discoveries, and insights with others, fostering a sense of unity and collaboration that transcends time and space.

As you continue your journey, remember that the Quantum Teleportation Kit is not only a powerful tool but also a reflection of the incredible potential that lies within each of us. By embracing the unknown and pushing the boundaries of human understanding, we can collectively advance our knowledge and pave the way for a brighter and more interconnected future.

Thank you for embarking on this journey, and welcome to the world of quantum teleportation. The future awaits, and the possibilities are truly limitless.

BIBLIOGRAPHY

"DREAM GIRL BAST Ario Bombacci." YouTube, uploaded by Transcender Starship, 19 Jan. 2017.

"Entropy Confusion - Sixty Symbols." YouTube, uploaded by Sixty Symbols, 3 Nov. 2014.

"Entropy Is NOT About Disorder." YouTube, uploaded by ZAP Physics, 10 Feb. 2019.

"Entropy Is Not Disorder: Micro-state Vs Macro-state." YouTube, uploaded by Physics Videos by Eugene Khutoryansky, 21 Jun. 2016.

"Great Eclipse at the Parthenon Hosted by Aneel Pandey." YouTube, uploaded by Transcender Starship, 22 Aug. 2017.

"Killing "Space-Time" Chapter 1-4: Rethinking General Relativity As 5 Dimensions of Physics - A Unifying Theory of Gravity." YouTube, uploaded by Chris "The Brain," 14 Jan. 2023. (See 33:24 / 1:01:59)

"Leonard Susskind | Lecture 2: Black Holes and the Holographic Principle." YouTube, uploaded by Mrtp, 28 Feb. 2015.

Pandey, Aneel. Psychedelic Yoga: Quantum Teleportation Techniques of the Bhagavad Gita. Published Independently, 2023.

Pandey, Aneel. The AI-Cryptography Collision: A New Banking

Emergency. Published Independently, 2023.

Nataraja Guru (Sanskrit to English Translation) and Nitya Chaitanya Yati (Explanatory Dialogue). The Bhagavad Gita: A Sublime Hymn of Yoga, 1981.

"Shannon Entropy and Information Gain." YouTube, uploaded by Serrano.Academy, 4 Nov. 2017.

Talbot, Michael. The Holographic Universe: The Revolutionary Theory of Reality. Harper Perennial, 2011.

Teitsworth, Scott. Krishna in the Sky with Diamonds: The Bhagavad Gita as Psychedelic Guide. Park Street Press, 2012.

"The Universe Isn't Locally Real – Quantum Entanglement." YouTube, uploaded by The Alchemist, 24 Jan. 2023.

"What Is Information Theory? (Information Entropy)." YouTube, uploaded by Art of the Problem, 14 Sept. 2012.

Wikipedia contributors. "Entropy (information theory)." Wikipedia, The Free Encyclopedia, 25 Mar. 2023. Web. 7 Apr. 2023.

Wikipedia contributors. "Introduction to entropy." Wikipedia, The Free Encyclopedia, 21 Feb. 2023. Web. 7 Apr. 2023.

www.ingramcontent.com/pod-product-compliance
Lightning Source LLC
Chambersburg PA
CBHW070644220526
45466CB00001B/290